Wilder Dwight Bancroft

The Phase Rule

Wilder Dwight Bancroft

The Phase Rule

ISBN/EAN: 9783743337398

Manufactured in Europe, USA, Canada, Australia, Japa

Cover: Foto ©ninafisch / pixelio.de

Manufactured and distributed by brebook publishing software (www.brebook.com)

Wilder Dwight Bancroft

The Phase Rule

CONTENTS

CHAPTER I
INTRODUCTION

Phase and component, 1. Phase Rule, 2. Theorem of Le **Chatelier,** 4. **Influence of gravity,** 5. Freedom from assumptions, 5.

ONE COMPONENT

CHAPTER II
GENERAL STATEMENT

Solid, liquid and vapor, 6. Addition of work and heat, **8. Changes at constant volume,** 10. Liquid and vapor, 11. Boiling-point, **12. Critical temperature,** 14. Solid and vapor, 16. Solid and liquid, 17. Vapor, **19. Liquid,** 20. Solid, 21. Labile states, 22.

CHAPTER III
WATER, SULFUR AND PHOSPHORUS

Water, 24. Sulfur, 28. Phosphorus, 32. Allotropy, 33.

TWO COMPONENTS

CHAPTER IV
ANHYDROUS SALT AND WATER

Compound and solution, 35. Solvent and solute, 36. **Potassium chloride and water,** 37. Cryohydrates, 38. Solubility and temperature, 41. **Critical temperature,** 44. Freezing-points, 46. Freezing-mixtures, **47. Divariant systems,** 50.

CHAPTER V
HYDRATED SALTS

Molecular and chemical compounds, 56. Sodium sulfate **and water, 57.** Pressure relations, 58. Efflorescence, 59. Heat of solution, **63. Dissociation** pressures, 65. Solubility curves, 66. Supersaturation, 68. Rational diagram, 70. Pressure curves for calcium chloride and water, 71. Solubility curves for calcium chloride and water, 75. Data, 77. Ammonia and ammonium bromide, 78. Solubility curves for ferric chloride and water, 79. Stable supersaturated solutions, 81. Data, 83. Thorium sulfate and water, 85.

CHAPTER VI
VOLATILE SOLUTES

Iodine and chlorine, 86. Solubility curves, 88. Data, 90. Gases and liquids, 91. Volatilization of solids, 92.

CHAPTER VII
TWO LIQUID PHASES

Naphthalene and water, 94. Types of partially miscible liquids, 96. Equilibrium between partially miscible liquids, 99. Boiling-points, 101. Solubility and temperature, 102. Consolute temperature, 103. Phenol, benzoic acid and salicylic acid with water, 105. Triethylamine and water, 106. Sulfur dioxide and water, 107. Concentration temperature diagram, 110. Data, 111. Hydrobromic acid and water, 112. Data, 114.

CHAPTER VIII
CONSOLUTE LIQUIDS

Naphthalene and phenanthrene, 116. Eutectic mixture, 117. Vapor pressure of consolute liquids, 118. Boiling-points, 120. Fractional distillation, 124. Solubility of consolute liquids, 126. Fusion and solubility curves, 128. Melted salts, 130. Solubility curves when compounds are formed, 131. Rule for number of constant freezing-points, 132. Two modifications of the solvent, 133. Mercuric and silver iodides, 134.

CHAPTER IX
SOLID SOLUTIONS

Freezing-points, 135. Solid and liquid solutions with the same composition, 137. Two sets of solid solutions, 137. Occlusion and adsorption, 139. Palladium and hydrogen, 139. Occlusion of gases by carbon, 140.

CHAPTER X
REVIEW

Typical nonvariant systems, 141. Alloys, 142. Metallic compounds, 143. Metallic solid solutions, 144. Amorphous antimony, 145.

THREE COMPONENTS

CHAPTER XI
GENERAL THEORY

Graphical methods, 146. Triangular diagram, 147. Simplest cases, 148. Theorem of van Rijn van Alkemade, 149. One compound, AC, 150. Two compounds, AC and AC, 152. Two compounds, AC and ABC, 153. Two compounds, AC and BC, 153. Potassium, sodium and lead nitrates, 154. Isotherms, 155. Two subdivisions, 156. Potassium chloride, potassium nitrate and water, 157. Disappearance of water, 158. "Break" when the solvent changes, 159. General form of isotherms when compounds can exist, 161.

CHAPTER XII

TWO SALTS AND WATER

Potassium sulfate, magnesium sulfate and water, 165. **Abnormal cryohydric temperature,** 166. Behavior of the double salts, 167. **Rule of Meyerhoffer,** 169. Diagram of van der Heide, 170. Application **of the theorem of van Rijn van** Alkemade, 170. Isotherms, 171. Data, 171. **Sodium sulfate, magnesium sulfate and water,** 172. Overlapping fields, 173. **Range of decomposition,** 174. Copper chloride, potassium chloride and **water, 175. Stability of** double salt, 177. Rule for cryohydric points, 178. **Potassium iodide, lead iodide and water,** 179. Calcium acetate, copper acetate **and water,** 179. **Maximum and minimum temperatures for double salts,** 180. **Reason for the rule,** 183.

CHAPTER XIII

PRESSURE CURVES

Sodium sulfate, sodium chloride and water, 184. **Efflorescence under water,** 186. Solubility of hydrated salt, 187. Calcium acetate, **copper acetate and water,** 188. Sodium sulfate, magnesium sulfate **and water, 190. Combined diagram,** 192. Copper chloride, potassium chloride **and water, 194. Efflorescence of hydrated double salts,** 195. Impossible formula **for a hydrated double salt,** 195. Behavior of lead potassium iodide, 196. **Intersection of dissociation curves,** 197. Ammonium chloride and lead oxide, **198.**

CHAPTER XIV

SOLID SOLUTIONS

Ether and rubber, 199. Starch and iodine, **199. Dyeing of silk, 200.** Occlusion by charcoal, 200.

CHAPTER XV

ISOTHERMS

Continuous addition of one component, 201. Salts do **not** react, 202. **Single series of solid solutions,** 203. Two series of solid solutions, 204. One double salt, 204. Two double salts, 205. One double salt and one series of solid solutions, 206. One double salt and two series of solid solutions, 207. Three series of solid solutions, 207. General rule, 208.

CHAPTER XVI

FRACTIONAL EVAPORATION

Isothermal evaporation with removal of crystals, 209. Salts do not react, 210. Single series of solid solutions, 210. Two series of solid solutions, 212. One double salt, 212. Behavior of lead potassium iodide, 213. Two double salts, 215. One double salt and one series of solid solutions, 216. One double salt and two series of solid solutions, 217. Three series of solid solutions, 218.

CHAPTER XVII

TWO VOLATILE COMPONENTS

Ferric chloride, hydrochloric acid and water, 219. Hydrated double salts, 221. Application of the theorem of van Rijn van Alkemade, 222. Forms of the isotherms, 223. Data, 224.

CHAPTER XVIII

COMPONENTS AND CONSTITUENTS

Dependent and independent variables, 226. Limiting conditions, 227. Water as monobasic and dibasic acid, 229. Adding degrees of freedom, 230. Classification independent of the chemical elements, 231. Systems with metathetical reactions, 231. Barium sulfate and sodium carbonate, 232. Mercuric sulfate and water, 232. Tartrates and racemates, 233. Passive resistance and time, 234. Semipermeable diaphragms, 235. Treatment of osmotic pressure, 236. Two different diaphragms, 236. Adiabatic diaphragms, 237.

CHAPTER XIX

TWO LIQUID PHASES

Second liquid phase independent of the third component, 238. Three liquid solutions, 239. Break in the isotherm, 240. Second liquid phase dependent on the third component, 240. Ammonium sulfate, alcohol and water, 241. Labile equilibrium, 242.

FOUR COMPONENTS

CHAPTER XX

GENERAL THEORY

Metathetical reactions, 243. Two pairs of salts and water, 243. Stable and instable pairs, 244. Number of components, 244. Data for inversion points, 245. Magnesium sulfate, potassium chloride and water, 246. Effect of adding magnesium chloride, 246. Data, 247. Pressure relations when the two magnesium sulfates are solid phases, 247. Indirect evidence, 248.

CHAPTER I

INTRODUCTION

The two expressions describing in a qualitative manner all states and changes of equilibrium are the Phase Rule and the Theorem of Le Chatelier. A phase is defined as a mass chemically and physically homogeneous[1] or as a mass of uniform concentration, the number of phases in a system being the number of different homogeneous masses or the number of masses of different concentration. In the case of water in equilibrium with its own vapor there is the liquid and the vapor phase, two in number. If there is a salt dissolved in the water there are still two phases, the liquid or solution phase and the vapor phase. If ice crystallizes, there is added a solid phase and the number becomes three. If, in addition, the dissolved substance separates in the solid form or as a second liquid layer, there will be four phases present, the vapor, liquid and two solid phases or the vapor, solid and two liquid phases as the case may be. Although the ice separates in many crystals, yet each is like every other in composition and density and taken together they constitute one phase. If the crystals were not alike as is the case with rhombic and monoclinic sulfur they would form as many phases as there were kinds of crystals, two in the example just cited, three if we have diamond, graphite and carbon. The components of a phase or system are defined as the substances of independently variable concentration in the phase or system under consideration. A component need not be a chemical compound, that is a substance described by the Theorem of Definite and Multiple Proportions, though this is usually the case. For instance, a mixture of propyl alcohol and water in such proportions that the percentage composition of the liquid is the same as that of the vapor might be treated as one component; but there is no advantage in this, as it is true for only one temperature and when

---

[1] Cf. Gibbs, Trans. Conn. Acad. **3**, 152 (1876).

there are no other components. The main point to be observed in determining the number of components in a given system is that each compound is not necessarily a component. Thus a hydrated salt is to be treated, when in equilibrium with the solution or vapor, as made up of salt and water and is not in itself a component. The same holds true of a double salt such as the double sulfates of copper and potassium. Here the components are the two single salts and water because the concentration of these three can be varied and they are sufficient to form all modifications which can exist. If one is treating calcium carbonate in equilibrium with calcium oxide and carbonic acid, there are only two components, calcium oxide and carbonic acid; for the calcium carbonate is merely a solid phase containing the two components. The fact that the two components unite to form a phase in definite proportions does not have anything to do with the matter. On the other hand it is not permissible to take calcium and oxygen as two of the actual components of this system because they are neither independent variables nor are they in equilibrium with the system.[1]

It has been shown by Gibbs[2] that the state of a phase is completely determined if the pressure and temperature together with the chemical potentials of its components be known. There is therefore an equation connecting these quantities which will describe the phase. For each other phase in equilibrium with the first there will be another equation containing the same variables.

There will thus be the same number of equations as there are phases, while the number of independent variables will equal the number of components plus the temperature and pressure. If the number of components be "n" the number of variables will be "$n + 2$." This is true only in case we are considering a system uninfluenced by gravity, electricity, distortion of the solid masses or capillary tensions because it is only when the effects due to these influences are removed that the values for the pressure, temperature and chemical potentials are uniform throughout the whole system. While we do not know the single equations referred to nor the chemical potentials of the components, it is possible to draw some

[1] Nernst, Theor. Chem. 482.
[2] Trans. Conn. Acad. 3, 152 (1876).

conclusions in respect to the possible number of states of equilibrium in any given case. Since the number of independent variables is always equal to "$n + 2$" by definition and the number of equations equals the number of phases, it follows that in a system of "$n + 2$" phases there will be as many theoretical equations as there are variables; in other words, that each of the variables has one value and one only for a given set of "$n + 2$" phases. A given combination of "$n + 2$" phases can exist at one temperature and one pressure only, the composition of the phases being also definitely determined. Such a system is called a nonvariant system, the temperature and pressure at which alone it can exist are known as the inversion temperature and pressure.[1] If there are only "$n + 1$" phases, the system is no longer completely defined and has one degree of freedom. It is therefore called a monovariant system. If we fix arbitrarily one of the variables, say the pressure or the temperature, the system is again entirely defined. The characteristics of the monovariant system are that for a given combination of phases there is for each temperature one pressure and one set of concentrations for which the system is in equilibrium; for each pressure, one temperature and one set of concentrations; for each set of concentrations, one pressure and one temperature. A system composed of "n" phases is called a divariant system. In it there are two variables which can be fixed arbitrarily before the system is completely defined. In such a system, for a given temperature, it is possible to have a series of pressures by changing the concentrations or a series of concentrations by changing the pressures. For a given pressure the temperatures can vary with changing concentrations and *vice-versa* while for definite concentrations there are similar relations between the pressures and temperatures. If instead of "n" phases the system contains $n - 1$, $n - 2$, etc., phases it is known as a trivariant, tetravariant system, etc. There are other terms in use, a monovariant system being called a "case of complete heterogeneous equilibrium" while a divariant system is known as a case of "incomplete heterogeneous equilibrium."[2] These phrases are unwieldy

---

[1] van 't Hoff, Études 142.
[2] Bakhuis Roozeboom, Recueil Trav. Pays-Bas 6, 266 (1887).

and unsatisfactory and must give way to the more rational nomenclature adopted here.[1]

By increasing the number of components and decreasing the number of phases it is possible to make a system with almost any degree of freedom; but, practically, a system ceases to be interesting from the qualitative point of view when it contains less than "n" phases because the possibilities are so numerous and so ill-defined. In the other direction, that of decreasing the components and increasing the phases, it is impossible to go.[2] Since "$n + 2$" phases constitute a nonvariant system which can be in equilibrium at one temperature and pressure only, a system of "$n + 3$" phases is most improbable and none such are known where there are no so-called passive resistances to change.[3] The discussion will therefore be limited to nonvariant, monovariant and divariant systems, starting with the number of components equal to one and increasing to four. Before beginning the study of the possible variations in equilibrium caused by changing the different variables and the number of phases, it is necessary to have some clue as to the direction of the change in equilibrium when there is an alteration in the system. This is given by the Theorem of Le Chatelier, which says: "Any change in the factors of equilibrium from outside is followed by a reverse change within the system."[4] If the external pressure is raised there is an increased formation of the component or phase occupying the lesser volume; if heat is added there is increased formation of the component or phase involving an absorption of heat; if the concentration of one component is increased in a given phase there is formation of the component or phase which involves a decrease in the concentration of the first component. In other words, the system in equilibrium tends to return to equilibrium by elimination of the disturbing element. It is now possible to take up distinct cases and see the

---

[1] This classification of systems into nonvariant, monovariant, divariant and so on, is due to Professor Trevor, who has used it for several years in his lectures.

[2] Gibbs, Trans. Conn. Acad. **3**, 153 (1876).

[3] Gibbs, Trans. Conn. Acad. **3**, 111 (1876).

[4] Comptes rendus, **99**, 786 (1884); Braun, Wied. Ann. **33**, 337 (1888); Cf. Duhem, Mécanique chimique, 152.

way in which the Phase Rule and the Theorem of Le Chatelier apply. It must be borne in mind that we are discussing only the states and changes of equilibrium which are due to the pressure, temperature and concentrations, and that the disturbing effects due to gravity, electricity, distortion of the solid masses and capillary tensions are eliminated. If this is the case the problem becomes simplified since the absolute mass of the phase has no effect on the equilibrium, because the concentration of a phase is not a function of its mass. A saturated solution remains saturated whether it is in contact with a small or a large amount of the solid. In the same way the equilibrium is not disturbed if the bulk of the solution be poured off. This would not be true if we were taking into account the effect due to gravity. Crystals at the bottom of a long tube filled with solution are under a greater pressure than if the liquid layer were but a few millimeters thick; and have different solubilities in the two cases. This is very noticeable in divariant systems of three components when there is a vapor phase in equilibrium with two liquid phases. An increase in the amount of the upper liquid layer produces a very distinct change in the mass of the lower liquid phase. This is a point which has been completely overlooked in the development of Nernst's Distribution Theorem.[1] The pressure of a gas in a tall cylinder is not strictly uniform owing to the influence of gravity. These effects as a rule are very small and may be neglected in most cases without danger. They can be reduced to a minimum by working with small quantities.

It will be noticed further that the classification of equilibria under the Phase Rule and of changes of equilibria under the Theorem of Le Chatelier is perfectly general and involves no assumptions as to the nature of matter or of the changes taking place. There is no need of assuming that matter is made up of discrete particles nor that it is continuous; there is even no need of assuming its existence or non-existence. It is immaterial whether there is or is not a distinction between "chemical" and "physical" reactions. It is simply a question of the relative number of independently variable components and phases in the one case and of the experimental data in regard to heat effects, densities and concentrations in the other.

[1] Zeit. phys. Chem. **8**, 110 (1891).

ONE COMPONENT

CHAPTER II

GENERAL STATEMENT

The most familiar example of a nonvariant system made up of one component is the equilibrium between solid, liquid and vapor, as in the system composed of ice, water and water vapor, or solid, melted and vaporized naphthalene. The application of the Phase Rule leads us to expect that a system of this type can be in equilibrium at only one temperature and one pressure. This is true experimentally, the temperature for water being about zero degrees Centigrade[1] and the pressure about four and a half millimeters of mercury. In the ordinary determinations of the inversion temperature in open vessels there are really two components instead of one, the substance under consideration and air, so that the values cited for water are not the temperature and pressure at which the three modifications of water would be in equilibrium if they alone were present. The effect of the air in changing the inversion temperature is due to its solubility and also to its pressure upon the solid and liquid phases. Since the effect of a pressure of one atmosphere is very slight and since the amount of air dissolved in water is not usually very great, the value of the inversion temperature as found in open vessels differs only slightly from the true value and for purposes of discussion the two may be considered identical in most cases. This is especially true if the determination is made by melting the solid in presence of its own vapor instead of freezing the liquid. In Table I. are the inversion temperatures and pressures of several substances. The first column of figures gives the temperature in Centigrade degrees; the second, the pressure in millimeters of mercury:

[1] More accurately $+ 0.0066°$ Gossens, Arch. Néerl. **20**, 449 (1886).

TABLE I

Bromine	−7.0°	·44.5
Ice	0.0	4.6
Benzene	5.3	35.4
Acetic acid	16.4	9.4
Naphthalene	79.2	9.
Iodine	114.2	90.
Camphor	175.	354.

If we start from any system in equilibrium it is possible, by adding or subtracting work or heat, to bring about changes in the relative masses of the phases and, under suitable conditions, the disappearance of one or more phases. The direction of these changes can be predicted from the Theorem of Le Chatelier if we know the densities and concentrations of the different phases and the sign of the heat effect when one phase increases at the expense of one of the others. The vapor of a substance is less dense than the liquid or solid modification at any temperature at which it can be in equilibrium with either of these.[1] The vapor of a substance is less dense than the liquid or solid modification. Increase of external pressure means therefore decrease of the vapor phase and *vice-versa*. Some substances are more and some are less dense in the solid than in the liquid state so that it is necessary to know the peculiarities of the system under consideration in order to tell which of these two phases is the more stable under increased pressure. The change of solid into liquid and of liquid into vapor always involves absorption of heat.

These preliminary experimental data being given we can now take up the changes in the relative masses of a nonvariant system, solid, liquid and vapor. If we have this system in a vessel closed by a movable piston so that it is possible to vary the pressure and volume we shall find, if we keep the system at the inversion temperature, and increase the external pressure that, as predicted by the Theorem of Le Chatelier, there will be formation of a system occupying a lesser volume. Some of the vapor will condense until the original pressure is restored, the volume of the vapor phase decreasing, that

---

[1] This would not be true for supercooled vapors near **the critical temperature**; but it holds for all cases of stable equilibrium.

of the liquid phase increasing, while the solid phase remains constant.

If the external pressure on the piston be kept continuously greater than the equilibrium pressure of the system, the condensation will continue until the vapor phase has disappeared and there is present the monovariant system, solid and liquid. If the external pressure be less than the equilibrium pressure, more vapor will form and the vapor phase will increase in volume at the expense of the liquid, driving back the piston till the equilibrium pressure is reached. If the external pressure be kept continuously less than the equilibrium pressure, the evaporation and expansion will go on until the liquid has disappeared and there is again a monovariant system, this time, solid and vapor. It might be thought that by bringing the solid and liquid phases into the two arms of a U-tube so that each is in contact with the vapor, evaporation would take place from both surfaces. In this case, by taking suitable proportions of solid and liquid, it would be possible to make the former disappear before the latter leaving the monovariant system, liquid and vapor. This will not happen, practically, because the conditions of the experiment are impossible. Owing to surface tension, the solid will be completely wetted by the liquid and will not be in direct contact with the vapor. This is a complication which has been ruled out expressly and we may conclude that, apart from the disturbing influences due to surface tension, it would be possible to pass by change of external pressure from the nonvariant system, solid, liquid and vapor to the monovariant system, liquid and vapor. In the purely theoretical case where the effect of surface tension is eliminated, it would be possible to have increase of both solid and liquid at the expense of the vapor phase when the external pressure is greater than the equilibrium pressure.

Instead of changing the external pressure, which is equivalent to adding work to or taking it from the system, it is possible to cause changes in equilibrium by adding or subtracting heat. This is done by bringing the system into contact with a body at another temperature than its own. For purposes of reference this outside body will be called the heat reservoir, and it may be at a higher or lower temperature than the system. Heat may be added or subtracted while the system is kept at constant pressure or at constant volume, the changes in the two cases being usually different. All instances where the pressure and volume change simultaneously may be resolved into the sum of two changes, one at constant pressure, the other at constant volume. So long as we have a nonvariant system before us, the addition of heat can bring about no change of temperature and it is necessary only to consider the changes in the relative and absolute masses of the phases. If the system is at a lower temperature than the heat reservoir and is kept at constant pressure, the solid will melt partially with absorption of heat, there being an increase in the liquid at the expense of the solid phase, the vapor phase remaining constant. This change will go on until the heat reservoir becomes of the same temperature as the system, that is until no more heat is added. If the heat reservoir be kept continuously at a higher temperature than the system, the change will continue until there is present the monovariant system, liquid and vapor. If the system is at a higher temperature than the heat reservoir and is kept at constant pressure, the reverse change will take place, the solid phase increasing at the expense of the liquid phase, the amount of vapor remaining constant. If the heat reservoir be kept continuously at a lower temperature than the system there will be formed eventually the monovariant system, solid and vapor. The change from liquid to solid is accompanied by an evolution of heat and is in accordance with the Theorem of Le Chatelier. It is to be noticed that the other monovariant system, solid and liquid, is not formed. It is not possible to pass at constant pressure from the nonvariant system, solid, liquid and vapor to the monovariant system, solid and liquid by addition or subtraction of heat.

If the nonvariant system be kept at constant volume and heat added continually, the changes and the resulting monovariant system will depend on the relative densities of the solid and liquid, and the relative masses of the three phases. If the solid is denser than the liquid—the usual case—the volume occupied by the liquid and solid will become larger as the solid melts and the volume of the vapor will become less. Under these circumstances there is increase of the liquid phase at the expense of the other two. If the quantity of solid is large and that of the vapor very small, the vapor phase will disappear first and the resulting monovariant system will be composed of a solid and a liquid phase. If the amount of solid is small and the volume of the vapor relatively large, the solid phase will be the first to vanish, leaving the monovariant system, liquid and vapor. If the solid is less dense than the liquid, as in the case of ice and water, the total volume of the solid and liquid phases will decrease with the melting of the solid. The liquid and vapor phases will both increase at the expense of the solid phase, and the resulting monovariant system will be made up of liquid and vapor irrespective of the original masses of the three phases. If the system is brought into contact with a heat reservoir kept continually at a lower temperature than its own, the reverse changes will take place. If the solid is more dense than the liquid, the solid and vapor phases will increase at the expense of the liquid phase, forming a system composed of a solid and a vapor phase regardless of the original volumes of the three phases. If the solid is less dense than the liquid, the solid phase will increase at the expense of the liquid and vapor phases, the resulting system being solid and vapor or solid and liquid as the volume of the vapor phase is large or small relatively to that of the liquid. It must be kept in mind that the main change in adding or subtracting heat is the conversion of solid into liquid and *vice versa*, and that the change in the vapor phase is a secondary one due to the difference in density of the solid and liquid. It is evident that the change to liquid and vapor on adding heat and to solid and vapor on subtracting it, is in accordance with the Theorem of Le Chatelier, the formation of liquid from solid being accompanied by absorption, the formation of solid from liquid by evolution of heat. This is not so obvious when the final state is solid and liquid on addition of heat. It is true that the conversion of vapor into liquid is accompanied by an evolution of heat, but this change is secondary, as has just been pointed out. There is very little vapor condensed and the sign of the heat effect is determined by the much larger change, as far as mass is concerned, of solid into liquid; there is actually an absorption of heat and the Theorem of Le Chatelier is again confirmed.

Having considered the changes in the nonvariant system due to addition and subtraction of heat and work, it is in order to treat in the same way the three monovariant systems, liquid and vapor, solid and vapor, liquid and solid. According to the Phase Rule these systems may exist over a series of temperatures and a series of pressures, bounded only by the appearance of new phases; but if the temperature is fixed, the pressure is also fixed and *vice versa*. This is the case experimentally and in Table II[1] are the values at different temperatures of the corresponding pressures for several examples of the monovariant system, liquid and vapor.[2] The fact of equilibrium between liquid and vapor being possible at different temperatures is familiar to everyone, but that the pressure is constant for constant temperature can be known only by quantitative measurements.

TABLE II

	Ether	Alcohol	Water	Iodbenzene	Mercury
0°	184.9	12.2	4.6		
10	291.8	23.8	9.1		
20	442.4	44.0	17.4		0.001
30	647.9	78.1	31.5	1.5	0.003
40	921.2	133.4	54.9	2.7	0.006
50	1276.1	219.8	92.0	4.8	0.013
60	1728.1	350.2	148.9	8.2	0.026
70	2273.9	540.9	233.3	13.6	0.050
80	2991.4	811.8	354.9	21.6	0.093
90	3839.7	1186.5	525.5	33.5	0.165
100	4859.0	1692.3	760.0	50.4	0.285

If the system, liquid and vapor, be subjected at constant temperature to an external pressure always greater than its own, there will be an increase in the denser or liquid phase, the vapor condensing until that phase has disappeared and there is present the divariant system, liquid. If the external pressure be kept constantly less than that of the system, and the temperature not allowed to change, the liquid will evaporate until there is left only the vapor phase. Water in an open vessel in a large room will evaporate completely because the concentration of water vapor in the room is less than that in

---

[1] The pressures are given in millimeters of mercury.
[2] For some interesting data cf. Barus, Phil. Mag. (5) **29**, 141 (1890).

equilibrium with liquid water. Under a bell jar the water evaporates until the equilibrium pressure is reached and then stops. If heat be added to the system kept at constant pressure, the vapor phase will increase at the expense of the liquid phase without rise of temperature till the latter has disappeared, a change which is accompanied by absorption of heat; if heat is subtracted under the same circumstances, the system will pass also without change of temperature, into the divariant system consisting of a liquid phase only. These two changes can take place at any temperature, but there is a form of the first one, occurring experimentally, which seems at first sight to be connected with a definite temperature. If a liquid is heated in an open vessel there seems to be no change beyond a rise of temperature until the boiling point, so-called, is reached, when the liquid distills off, the temperature remaining constant. In the first place, the liquid evaporates at all the intermediate temperatures, though this is not noticed, the quantity being usually small under the conditions of the experiment. In the second place, there is air in the vessel, so that we are no longer considering a system made up of one component. For all that, it is better to treat the subject here rather than later, since the air is really only a disturbing element and not an integral part of the system. The temperature between the system and the heat reservoir—the Bunsen burner, for instance—is so great that the liquid vaporizes faster than it can diffuse out of the vessel.[1] The system acts to a certain extent as if it were receiving heat while at constant volume, the temperature rising and the vapor pressure increasing. The vapor phase in the vessel, being air at atmospheric pressure plus the vapor of the liquid, is at a higher pressure than the external, atmospheric pressure and diffuses out against it. The vapor phase becomes ever richer in the vapor of the liquid and poorer in air till all the air has been driven out and there is a true monovariant system, liquid and vapor. This occurs when the vapor pressure of the liquid is equal to the atmospheric pressure. If the atmospheric pressure does not change, we have a monovariant system at constant pressure and the temperature will remain constant until

---

[1] In the spheroidal state, on the other hand, the evaporation is so rapid that water does not rise to its boiling point. Cf. Ramsay and Young, Phil. Trans. **175**, 47 (1884).

the liquid phase has **disappeared. If the** atmospheric pressure be changed in any way the **phenomena of** boiling will occur at the temperature at which **the vapor pressure is equal** to the modified atmospheric pressure. **If the external pressure be** changed sufficiently **the liquid may be made to boil** at any temperature at which it can **exist in equilibrium with** the vapor, from the inversion to **the** critical **temperature.**[1] **If,** instead of allowing the liquid to distill **off, the flask is connected** with a reverse cooler so that the condensed **liquid runs back, there is** only vapor of the liquid in the flask, while **outside there is air, and it** may be asked why the air does not diffuse **into** the flask. **There is** a tendency for it to do this but the air is **carried back by the** constant current of vapor streaming up. **There will be** a continual decrease in concentration **of the air** from outside **to the** point where the vapor **pressure of the liquid is equal to the atmospheric pressure.** If the **cooling is sudden so that the liquid condenses** all at one place, **there will be an abrupt change from all air to all vapor.** If the condensation **takes place along quite an interval, as usually happens with glass condensers, there will be a gradual transition, the amount of air decreasing and that of vapor increasing as** one **approaches the liquid. In other words, when the pressure of** a saturated vapor is kept **equal to the external pressure, the vapor is impermeable to other vapors and gases.** The determination of the boiling point in a flask with a **reverse cooler is open to** one objection due to the influence of gravity. **If the liquid is boiling regularly and** the flame underneath be turned **up, the rate of vaporization will** be increased and the vapor **will condense at some higher point in the cooler.** This will be accompanied **by a slight rise of** temperature which **if not** taken into **account may produce errors when using the Beckmann boiling-point apparatus. The way to avoid this is to regulate the gas pressure or the cooling in the condenser so that the precipitation may take place always at the same point.**

If heat be added to a system, **liquid and vapor, kept at constant volume there will be an increase of vapor at the expense of the liquid, that change absorbing heat, and the vapor pressure will in-**

---

[1] A definition **of critical temperature is given on the next page.**

crease. The system can not be in equilibrium with the new vapor pressure at the same temperature and the temperature rises with absorption of heat. This will continue in many cases until the liquid has disappeared and there is present the divariant system, vapor. If the volume of the liquid is large relatively to that of the vapor, the course of events is somewhat different. On adding heat there will be an increase of temperature and pressure until a definite temperature and pressure has been reached when the surface of the liquid which has been fairly clearly defined hitherto, suddenly billows up and disappears, the contents of the vessel becoming homogeneous. On cooling the reverse change takes place, a tumultuous commotion in the tube and the formation anew of two phases. Whether the contents of the vessel are vapor or liquid is impossible to determine because the two are identical. The temperature and pressure at which this phenomenon takes place are known as the critical temperature and pressure. If the system be heated above this temperature and allowed to expand, it passes without discontinuity into what is certainly the gaseous state; if the pressure is increased and the system allowed to cool it passes also without discontinuity into what is unmistakably the liquid phase.[1]

TABLE III

Hydrogen	$-234.5°$	20.
Nitrogen	-146.0	35.0
Oxygen	-118.8	50.8
Carbonic acid	31.0	77.0
Ether	194.4	35.6
Acetone	234.4	60.
Alcohol	243.6	62.8
Chloroform	260.0	54.9
Benzene	288.5	47.9
Water	365.	200.5

[1] Andrews, Phil. Trans. **2**, 575 (1869); Sajontchewsky, Beibl. **3**, 741 (1879); Nadeshdin, Ibid. **8**, 721 (1884); Cailletet and Colardeau, Comptes rendus, **112**, 563 (1891); Altschul, Zeit. phys. Chem. **11**, 577 (1893); Galitzine, Wied. Ann. **50**, 521 (1893); van der Waals. "Die Kontinuität des gasförmigen und flüssigen zustandes." Leipzig (1881); Cf. Landolt and Börnstein's Tabellen, 91.

At the critical temperature the liquid and vapor have the same density. In Table III are the critical temperatures and pressures of several substances.[1] The pressures are given in atmospheres.

This difference in behavior of a substance above and below the critical temperature renders it possible to make a distinction between a gas and a vapor, a vapor being a gas below its critical temperature and a gas a vapor above that temperature. A vapor can be condensed to a liquid by increase of pressure without change of temperature while a gas cannot be; it must be cooled as well. This distinction has been of great importance in the attempts to liquefy substances hitherto known only in the gaseous state. It had been found that certain substances were liquefied by subjecting them to great pressures and it was assumed that with sufficiently high pressure any gas could be liquefied. All attempts in this direction proved futile with such gases as nitrogen, oxygen and hydrogen. After the experiments of Andrews on carbonic acid the reason for this failure became clear. By cooling below the critical temperature and using high pressures it has been possible to liquefy all the so-called permanent gases.[2]

Starting from the nonvariant system, solid, liquid and vapor, we may cause the liquid phase to disappear leaving the monovariant system, solid and vapor. For this system as for all monovariant systems the statement holds true, that for each temperature there is a definite pressure under which the system is in equilibrium; but this is not very easy to show from direct measurements owing to the small vapor pressures of most solids.

In Table IV are the determinations of the vapor pressures of ice, benzene, acetic acid and camphor at different temperatures. The pressures are given in millimeters of mercury.

---

[1] Heilborn, Zeit. phys. Chem. **7**, 601 (1891).

[2] Faraday, Phil. Trans. **113**, 160 (1823); **145**, 1 (1845); Natterer, Sitzungsber. Akad. Wiss. Wien, **5**, 351 (1850); **6**, 557 (1851); **12**, 199 (1854); Cailletet, Ann. chim. phys. (5) **15**, 132 (1878); Pictet Ibid. **13**, 145 (1878); Wroblewski and Olszewski, Wied. Ann. **20**, 243 (1883); Olszewski, Phil. Mag. (5) **39**, 237; **40**, 202 (1895).

TABLE IV

Water		Benzene		Acetic Acid		Camphor	
Temp.	Pressure	Temp.	Pressure	Temp.	Pressure	Temp.	Pressure
$-10°$	2.03	0.°	24.42	1.85°	2.35	20.°	1.0
-8	2.37	1.	26.18	6.41	3.75	35.	1.8
-6	2.81	2.	28.08	9.16	4.70	62.4	6.4
-4	3.33	3.	30.03	12.10	6.05	78.4	9.5
-3	3.62	4.	32.32	13.30	6.75	100.	22.6
-2	3.94	5.	34.65	14.30	7.20	132.	78.1
-1	4.28	5.3	35.41	15.80	8.85	154.	188.8
0	4.64			16.41	9.45	175.	354.

The most interesting experiments showing the qualitative existence of a vapor pressure are those of Demarçay,[1] of Hallock[2] and of Spring.[3] Demarçay placed the metals to be examined in a tube connected with a Sprengel pump. Heating one end of the tube and cooling the other he sublimed cadmium at 160°, zinc at 184°, antimony and bismuth at 292°, lead and tin at 360°. Hallock found that sulfur combined with copper and other metals even when separated from them by a long tube filled with plugs of cotton wool to prevent convention currents. Spring placed zinc and copper near together but not in contact and observed the formation of brass at temperatures well below the melting point of the more fusible metal.

If work be subtracted from the system by keeping the external pressure always a little less than the equilibrium pressure there will be increase of the vapor, the less dense phase, at the expense of the solid phase; the solid will evaporate until there is formed the divariant system, vapor. The evaporation of ice in a cold, dry room is a well-known phenomenon. If work be added to the system there will be condensation of vapor until there is only solid present.

If heat be added to or taken from the system while it remains at constant pressure, the solid will evaporate or the vapor condense as the case may be without change of temperature. The change when

[1] Comptes rendus, **95**, 183 (1882).
[2] Am. Jour. Sci. (3) **37**, 402 1889.
[3] Zeit. phys. Chem. **15**, 76 (1894).

heat is added is in the opposite direction **to that** taking place on addition of work. If heat be added and the system kept at constant volume there will be increase of temperature and of pressure with formation of the vapor phase at the expense **of** the solid. **If there is relatively little** of the solid phase, the transition **will be to the divariant system, vapor;** otherwise **to the nonvariant system, solid, liquid and vapor.** It was found during the discussion of the equilibrium between liquid and vapor that, in the presence **of air, the system behaved to a certain** extent as if it were kept at constant **volume until the pressure of** the vapor equalled the external pressure. **This is the case in the** equilibrium between solid and vapor. **If ammonium chloride** be heated in a test tube, the **temperature will rise until the pressure of** the ammonium **chloride vapor is equal to the** barometric pressure and then the solid **sublimes without further change** of temperature. By altering the external **pressure the sublimation** temperature will change **just as the boiling temperature changes.** If the sublimation **temperature be lower than the inversion** temperature **the** solid **will sublime; if it be higher the solid will melt.** Ammonium chloride **melts when heated under sufficient pressure;** ice sublimes if **the external pressure is four millimeters of mercury or** less.[1]

The monovariant system, solid and liquid, is characterized **like the** others by being fully determined when the pressure **or temperature** is fixed or **the** density of either of the phases. **Increase of external pressure produces** in some cases disappearance **of the liquid, in others of** the solid phase. **In all** instances, as predicted **by the Theorem of** Le Chatelier, it is **the** less dense phase **which disappears, the solid** in the case of ice and water, the liquid **with most substances. Decreasing** the external pressure produces **the reverse change. Adding** heat causes disappearance **of the solid phase without change of temperature** if the system be **kept at constant pressure. If the system** be kept at constant volume the **addition of heat** causes transition to the nonvariant system, solid, **liquid** and vapor with **rise** of temperature and fall of pressure if the solid is less dense than liquid, transition to the divariant **system,** liquid, with rise of temperature and

---

[1] Ramsay and Young. Phil. Trans. **175,** 37 (1884).

pressure if the solid is denser than the liquid. Taking heat from the system leads in the first case to the solid phase, in the second to the nonvariant system. The temperatures at which solid and liquid can be in equilibrium are lower than the inversion temperature if the solid is less dense than the liquid and are higher if the contrary is true. The freezing point of a substance being defined as the temperature at which solid and liquid can be in equilibrium, it is lowered by increased pressure if the liquid is denser than the solid,[1] otherwise it is raised.[2] In all cases it is found experimentally that there is only one temperature for each pressure at which the solid and liquid are in equilibrium. In Table V are some of the experimental data on the change of the freezing point with the pressure.[3] The pressures are given in atmospheres; under the heading temperature are given the changes in temperature, the melting point at atmospheric pressure being taken as the standard.

TABLE V

Water		Naphthalene		Naphthylamine	
Pressure	Temperature	Pressure	Temperature	Pressure	Temperature
0.06	+0.0066°				
8.1	−0.059	8.0	+0.282	8.0	+0.105
16.8	−0.129	12.0	+0.405	12.0	+0.180

It is unknown whether there is a temperature and pressure at which solid and liquid become indistinguishable as is the case for liquid and vapor; but it is probable that there is such a point and that the systems, solid and liquid, liquid and vapor, can exist only between two critical temperatures.[4]

---

[1] W. Thomson, Phil. Mag. (3) **37**, 123 (1850); Dewar, Proc. Roy. Soc. **30**, 533 (1880).

[2] Battelli, Atti del R. Ist. Ven. (3) **3** (1886); Amagat, Comptes rendus, **105**, 165 (1887).

[3] Cf. Ostwald, Lehrbuch I, 1013–1015.

[4] Cf. Voigt, Kompendium der Physik I, 583. If it were not for the capillary phenomena this would occur when solid and liquid had the same density. Experimentally, this is not the case. Cf. **Damien**, Comptes rendus, **112**, 785 (1891).

There remain only the divariant systems, solid, liquid, and vapor to consider. Vapors and gases fill the vessel in which they are contained uniformly, barring the influence of gravity, and exert a pressure on the walls. For any temperature there is possible any pressure so long as no new phase appears, and for any pressure, any temperature. Increase of external pressure produces decrease of volume and *vice versa*. At constant pressure the addition of heat produces an increase of volume and a rise of temperature, both changes involving an absorption of heat. At constant volume addition of heat causes rise of temperature and of pressure. It is to be noticed in all cases where there is one phase and one component, that keeping the mass and the volume constant is the same as fixing the concentration and there is then only one degree of freedom left, one independent variable. For each temperature there will be only one pressure possible for a given concentration. It may be noted, in passing, as an experimental fact that the same concentrations of two gases, expressed in grams per liter for instance, do not give the same pressure at the same temperature. The explanation of this together with the mathematical statement of it in the form of the Gas Theorems belongs under the head of Quantitative Equilibrium and will not be taken up in this book. We have found heretofore that addition of heat to the system kept at constant pressure has produced no change in temperature and the behavior of a vapor seems to be an exception. This is only apparent, however. In the previous cases there have been always two or more phases, and there has been a transference of matter from one phase to another without change of concentration in any of the phases. This can not happen where there is only one phase and the volume changes, which always occur when heat is added to a system kept at constant pressure, produce a change in the concentrations. If the pressure is fixed, there is only one independent variable ; when the concentrations remain constant the temperature remains constant, and when the concentrations change the temperature changes. Addition of heat to a system kept at constant pressure produces no change in temperature if the concentrations of the phases remain unchanged, a rise of temperature if the concentrations change. This is not yet entirely satisfactory because it leaves undecided the question under what circumstances the concentrations do or do not change. In a nonvariant system there is no change of pressure, temperature or concentrations. In a monovariant system the temperature and the concentrations can not change when the pressure remains constant. No change of temperature is produced when heat is added to a nonvariant or a monovariant system kept at constant pressure; in all other cases there is a rise of temperature. It may be interesting to consider what changes in pressure and volume will take place on adding heat, the temperature remaining constant. Since all systems expand on addition of heat when there is no change of phase, there will be always an increase of volume which carries with it, by the Theorem of Le Chatelier, a decrease in pressure.

Liquids differ from gases and vapors in that they do not necessarily fill the whole of the vessel which contains them, and in that they have a form of their own. Since liquids fill the lower part of the containing vessel completely and are bounded on the upper side by a horizontal, nearly plane surface, it might be thought that they had no definite shape of their own. This behavior is due to the influence of gravity, and when this is eliminated, by suspending the liquid in a fluid medium of approximately the same density, we perceive that the spherical shape is the true form, characteristic of all liquids. For each temperature there can be a series of pressures at which the liquid can exist and for each pressure a series of temperatures, limited in both cases only by the appearance of a new phase. While the change of volume with change of pressure is fairly large with gases and vapors[1] it is very small with liquids, so small in fact that it can be shown only by careful quantitative measurements.[2]

The solid phase is characterized by rigidity and elasticity; two mutually exclusive properties. By its rigidity it resists deformation and is therefore not dependent on the vessel in which it is contained

---

[1] Cf. v. Lang, Theor. Physik. 650; Ostwald, Lehrbuch I, 139-159.

[2] Oersted, Pogg. Ann. **9,** 603 (1827); Pagliani and Vicentini, Beibl. **8,** 794 (1884); Röntgen and Schneider, Wied. Ann. **29,** 165 (1886); Schumann, Ibid **31,** 14 (1887); Boguski, Zeit. phys. Chem. **2,** 126 (1888); Amagat, Jour. de Phys. (2) **8,** 197 (1889).

for its shape. By its elasticity it returns to its original shape after having been subjected to a deforming stress. These properties are relative only and in some cases almost imperceptible. A steel spring goes back to its original form even after having been very much compressed; a piece of putty remains in the new shape. The rigidity varies very much also. All solids flow a little if left in a state of strain. The behavior of sealing-wax is well known. Car-axles become changed in structure through the continual jolting. Spring[1] has shown that, if two clean metal surfaces be brought into intimate contact, they unite by diffusion, and, in many cases, the bars thus formed can be placed in a lathe with one end free and have shavings turned from them without breaking. When the sticks were broken by twisting, the fracture did not come at the junction but usually across it. While a copper wire bends without breaking, showing a power of internal readjustment, a stick of bismuth is so brittle that it fractures under a very slight strain. Whether the crystalline structure or tendency to assume definite shapes bounded by plane surfaces is a characteristic of solids is a doubtful question. Solids certainly occur in what is known as the amorphous state in which no signs of a crystalline structure can be detected by any means at our disposal. Nernst[2] has made the suggestion that the solid is really present in very minute crystals; but he offers little evidence in behalf of this view, and it is not generally accepted.

As predicted by the Phase Rule, the solid phase can exist at a series of temperatures and for each temperature at a series of pressures limited only by the appearance of new phases. Increase of external pressure causes diminution of volume, and may cause appearance of the liquid phase if the solid is less dense than the liquid. Decrease of external pressure is accompanied by expansion and eventually by formation of one of the monovariant systems, solid and liquid or solid and vapor, as the case may be. Addition of heat to a solid kept under constant pressure produces increase of volume and temperature with eventual formation of solid and liquid or solid and vapor, depending on the nature of the substance and the initial pressure and temperature. If the system is kept at con-

[1] Zeit. phys. Chem. **15**, 70 (1894).
[2] Theor. Chem. 65.

stant volume addition of heat causes increase of temperature and pressure and transition to the monovariant system, solid and liquid.

A system is said to be in stable equilibrium when the addition of any modification of any of the components produces a change in the system proportional to the quantity of substance added.[1] The equilibria which we have considered so far have all been of this type. If the solid phase be added, for instance, to the monovariant system, liquid and vapor, the solid will melt except at the inversion temperature, when it is in equilibrium. The changes in either case will be proportional to the quantity of solid added.

It is possible to have a system in equilibrium with respect to the phases then co-existing which shall not be in equilibrium when brought in contact with some other modification of the components. Such a system is said to be in labile equilibrium because the equilibrium though stable as regards the phases already present is instable with respect to some other phase.[2] The most familiar example of this is the supersaturated solution of sodium sulfate which, if left to itself, will remain unchanged for months, perhaps years. If a crystal of the hydrated salt be thrown in, there is a sudden crystallization and the quantity of the new phase formed bears no relation to the amount of salt added to start the reaction. These labile equilibria occur in all systems, and we will take up first the supercooled vapors. If a vapor be cooled at constant volume, there will be reached a temperature and pressure at which there should be formation of liquid, and this usually takes place. It is possible, however, by careful cooling, especially if there be no dust present, to pass this point without the liquid phase being formed.[3] This equilibrium is now labile, for the addition of the smallest quantity of liquid produces a sudden condensation which ceases only when the vapor pressure characteristic of the monovariant system at that temperature has been reached. A second form of labile equilibrium, that of

---

[1] Gibbs, Trans. Conn. Acad. **3**, 455 (1878); Meyerhoffer, Die Phasenregel, 10.

[2] It is a step backward to class a supersaturated solution and a mixture of hydrogen and oxygen under the same head as Duhem (Mécanique chimique, 158) has done. Addition of water as vapor or liquid to the hydrogen and oxygen mixture produces no change It is not a case of labile equilibrium.

[3] R. v. Helmholtz, Wied. Ann. **27**, 521 (1886).

a superheated liquid, **is even harder to** observe experimentally. Donny[1] has shown that it is possible by unequal heating to raise the temperature of liquid water to **138°** without its boiling. This **can be** done only by keeping the surface **of the** water cool so that **the** superheated liquid is not in contact with **the vapor.** Dufour[2] attained the **same result by suspending** drops **of water in a** mixture of **linseed oil and oil of cloves having the same density as water. A temperature of** 175° **was reached** in this way. **The third volume of van der Waal's**[3] **is** a mathematical fiction and there seems **to be no good** experimental ground for assuming that the vapor **pressures of superheated** liquids and supercooled vapors are parts of **the same curve.** It is possible to obtain another instance of labile equilibrium **by** cooling a liquid in the absence of dust **below the inversion temperature.** With most liquids it is possible **to supercool the liquid** a degree or so and with some the supercooling **can be carried much farther.**[4] The reverse case of a solid **heated above its melting point** has never been realized.[5] **The third monovariant system, solid and** vapor has not been studied with **the same care as the other two, and** I am not able to cite **any quantitative example of a supercooled** vapor though it can hardly **be an uncommon phenomenon.**[6] The "hanging" of mercury in barometer tubes[7] **is an instance of a** vapor phase not being formed on decrease **of pressure, though this** is somewhat complicated by surface tension phenomena. **In all the cases of** labile equilibrium the presence of the **smallest quantity of** the phase in respect to which the system **is instable brings about a** change bearing **no** relation in its extent to the quantity **of that phase.**

[1] Pogg. Ann. **67**, 562 (1846); Cf. **Gernez, Comptes rendus, 86, 472; 87,** 1549 (1876).

[2] Ann. Chim. Phys. (3) **68, 370 (1863).**

[3] Ostwald, Lehrbuch I, 299.

[4] Schrotter, Sitzungsber. Akad. **Wiss. Wien, 10,** 527 (1853); Schröder, Liebig's Annalen **109,** 45 (1859).

[5] In the case cited by Ostwald (Lehrbuch **I, 994), there are two components.**

[6] Cf. Lehmann, Molekularphysik, II, 581.

[7] Moser, Pogg. Ann. **160,** 138 (1877).

CHAPTER III

WATER, SULFUR AND PHOSPHORUS

The general results, which have been enumerated in regard to equilibrium, will be grasped more easily if they are represented graphically. In Fig. 1 are shown the limiting values of pressure and temperature for water in its different modifications. The ordinates are pressures and the abscissae temperatures; the drawing is not to scale.

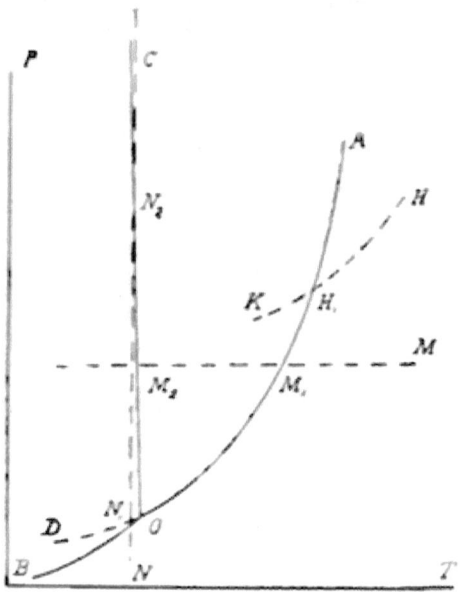

Fig. 1.

The curve OA is the vaporization[1] curve, showing the pressures and temperatures at which liquid and vapor can coexist, and it can be seen from the diagram that for each temperature there can be

---

[1] Bakhuis Roozeboom, Recueil. Trav. Pays-Bas **6, 280** (1887).

only one pressure and for each pressure only one temperature at which these two phases can be in equilibrium. The curve terminates at O because the solid phase appears; at the other end it is limited by the critical temperature, of 365°, and the critical pressure of 200 atmospheres, the difference between the two phases disappearing.[1] The curve OB is the sublimation curve, the solid and vapor phase being in equilibrium. The curve OC represents the equilibrium between solid and liquid. In this case it slants to the left because ice is less dense than liquid water.

In the corresponding diagram for naphthalene, it would slant to the right because the solid is more dense than the liquid and the freezing point is therefore raised by pressure. We do not know whether there is an upper limit beyond which this curve can not extend. The curve OD is the continuation of AO and represents the labile equilibrium between water and vapor. It will be seen from the diagram that this curve lies above the curve for ice and vapor[1] and is therefore instable with respect to it, as vapor will distill from the water and condense on the ice until the former disappears. This is the easiest way of making the facts intelligible though it is not really accurate; because the change is not one of distillation from a place of high pressure to one of low pressure, but a direct change of liquid into solid, as appears from the velocity of the reaction. In the present state of our knowledge we can express instability only in pressure and concentration differences although that is obviously a very incomplete statement of things as they are. The three curves AO, BO and CO meet at the point O, the inversion point, if we neglect the effects due to surface tension. In an actual system it is quite possible that this is not the case[2] though this has never been shown experimentally. If the three curves do not meet at a point the intersection of AO and CO will give the inversion temperature and pressure. Since the nonvariant system can exist at the point O and the monovariant systems each along one of the curves,[4] it fol-

---

[1] Cailletet and Colardeau, Comptes rendus, **112**, 1170 (1891).
[2] Cf. Ramsay and Young, Phil. Trans. **175**, II, 461 (1884); Ferche, Wied. Ann. **44**, 265 (1891).
[3] Wald, Zeit. phys. Chem. **7**, 514 (1891).
[4] Usually known as boundary curves.

lows that the divariant systems are in stable equilibrium in the fields bounded by the curves. Vapor can exist at any temperature and pressure in the field AOB, liquid in the field AOC, and solid in the field COB. The diagram summarizes also what we have learned about the effect of addition or subtraction of heat and work. If we start at some point M in the field AOB where the pressure is greater than the inversion pressure and subtract heat, keeping the pressure constant, the temperature will fall until the point M_1 is reached, when the liquid phase appears and the temperature remains constant until the whole of the vapor phase has disappeared. The temperature will fall again until at the point M_2 the solid phase appears and the temperature begins to decrease once more only when the liquid phase has disappeared. If we start from the point H and subtract heat while keeping the system at constant volume, the pressure and temperature will change in a way depending on the equation of condition for the vapor. This is indicated, without any attempt at accuracy, by the line HH_1. At H_1 the vapor is in equilibrium with the liquid; but, if the cooling is done carefully, it is possible to prevent condensation and to realize a small portion of the dotted curve H_1K when the vapor is in labile equilibrium. Ordinarily condensation takes place at H_1 and there is formed the monovariant system, liquid and vapor. The boundary curves are curves for equilibrium at constant volume so long as there is a monovariant system present. On further subtraction of heat, the pressure and temperature fall, the corresponding values always lying on the curve H_1O. At O the solid phase appears and both temperature and pressure remain constant until one of the phases has disappeared. Usually this will be the liquid phase and subsequent pressures and temperatures are represented by the line OB. In the case of water the vapor phase may disappear if its volume is small in comparison with that of the liquid, whereupon further subtraction of heat will cause the system to pass along the curve OC, the temperature falling and the pressure increasing. If the solid and vapor are cooled with due precautions, the solid phase may not appear at O and the pressures and temperatures of supercooled water are observed, represented by the dotted line OD. If we start from a third point N in the field AOB and increase the external pressure, keeping the temperature constant, the diagram tells us that the pressure of the system may increase indefinitely, remaining constant, however, while the vapor condenses to ice at N_1 and while the ice melts to water at N_2. From this brief review it is clear that the diagram gives a concise, intelligible summary of all the facts and, in treating other cases, it will not be necessary to go over the ground twice as has been done this time; it will be sufficient to start from a diagram and study that. It is to be noticed that the treatment has been entirely general and the results are independent of the equations of condition describing or failing to describe the three phases. It is immaterial whether water vapor dissociates into hydrogen or oxygen at the temperature of the experiment or not, provided the composition of each phase may always be represented empirically by the formula, H_2O. If the system is in a state of reversible equilibrium, it is indifferent whether there is dissociation, association or neither in the vapor phase; if the system is not in equilibrium it can not, of course, be represented in a diagram which gives equilibrium pressures and temperatures only.[1] It is not permissible to add an excess of either hydrogen or oxygen because the system would then contain two components instead of one. One word on another point may not be superfluous. In the previous discussion the addition of work and the addition of heat have been treated as if they were two independent processes. This is not true. When work is added to or, more properly, done upon the system there is a heat effect produced in all except adiabatic changes. In like manner when heat is added to the system, there is always addition or subtraction of work except in the one case when the volume is kept constant. It has seemed advisable to concentrate the attention first on the one effect and then on the other, ignoring the simultaneous manifestations of energy in other forms. These unconsidered energy changes are referred to in the provisions that the temperature be kept constant when the addition or subtraction of work is under consideration and that the pressure be kept constant when the effects produced by the addition or subtraction of heat are the interesting changes.

[1] Nernst (Theor. Chem. 487) does not seem to be very clear on this point. He speaks of the mixture of hydrogen and oxygen as being in labile equilibrium, and in the same paragraph (with a reference to p. 532) of its not being in equilibrium at all.

Water is a compound which can exist in only three forms, solid, liquid and vapor.[1] While there are no substances known which form two liquid phases, there are quite a number which can occur in two or more solid modifications. This brings in other inversion temperatures, and we will take up first the case of sulfur which exists certainly in the rhombic and monoclinic forms and possibly in other modifications. In Fig. 2 is a graphical representation of our knowledge in respect to the coexisting phases of sulfur; the abscissae denote temperatures and the ordinates pressure as before. The diagram is not to scale.

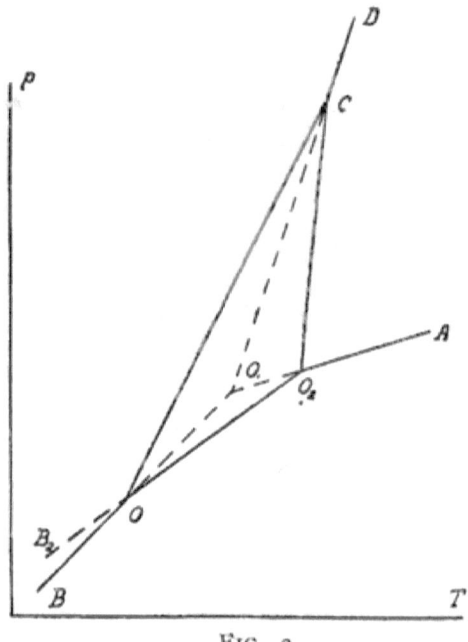

FIG. 2.

At ordinary temperatures the rhombic crystals are the more stable and the curve O_1B is the sublimation curve for rhombic sulfur. O_1A is the curve for the monovariant system, liquid and vapor, while O_1C represents the equilibrium between rhombic and liquid sulfur. In this there is nothing different from the behavior of water

---

[1] Cf. however, Prendel, Zeit. Kryst. **23**, 76 (1894).

except that the line O_1C slants off to the right. The inversion temperature at which rhombic and liquid sulfur can exist in presence of vapor is 114.5°. This is not the only inversion temperature. The curve O_2B_1 represents the equilibrium between monoclinic sulfur and vapor, O_1A that between liquid sulfur and vapor, and O_2C that between monoclinic and liquid sulfur. The temperature of O_2 is 120°. The curves O_1B and O_2B_1 intersect at O, giving a new triple point with the three phases, rhombic and monoclinic sulfur and vapor. From this point starts the curve OC showing the conditions of equilibrium in the monovariant system, rhombic and monoclinic sulfur. The temperature at which this third nonvariant system exists is 95.4°. The pressure has not been determined. Not all of these curves are stable. The curves BO_1 for rhombic sulfur and vapor ceases to be stable at O, and the remainder of the curve OO_1 represents a labile equilibrium. The pressures are higher than those for the vapor in equilibrium with monoclinic sulfur, shown by OO_2, and addition of monoclinic sulfur causes a complete conversion of the rhombic sulfur into the more stable form. If monoclinic sulfur is not added, the change does not take place readily and the curve can be followed to the melting point of rhombic sulfur at 114.5°.[1] The liquid sulfur has a higher vapor pressure than the monoclinic sulfur at that temperature and the curve O_1A is therefore one of labile equilibrium as far as O_2. The curve O_2C for rhombic and liquid sulfur is instable in respect to the monoclinic form and has not been studied. Monoclinic sulfur can exist in stable equilibrium with the vapor of sulfur from the melting point at O_2, 120°, to the other inversion point at O, 95.4°. Below this temperature, rhombic sulfur has the lesser vapor pressure and is the more stable form; above it, this is reversed, while at the temperature and pressure represented by O, and at this temperature and pressure only can rhombic and monoclinic sulfur exist in stable equilibrium with each other and with sulfur vapor. The work on this point has been done by Reicher[2] under van 't Hoff's direction. If there is no vapor of sulfur present we have a monovariant system instead of a non-

---

[1] Brodie, Phil. Mag. (4) **7**, 439 (1894).
[2] Recueil Trav. Pays-Bas, **2**, 246 (1883); Zeit. Kryst. **8**, 593 (1884).

variant one and there can therefore be equilibrium between monoclinic and rhombic sulfur at other temperatures provided the system is under proper pressure. This is shown graphically by the curve OC,[1] a few points of which have been determined by Reicher. Since the rhombic sulfur is the denser form, increase of pressure must raise the temperature at which it can be in equilibrium with monoclinic sulfur. Reicher found that for a pressure of four atmospheres the corresponding temperature was 95.6°, and for a pressure of a little less than sixteen atmospheres 96.2°, a change of 0.6° for twelve atmospheres.[2] Although these results agree both qualitatively and quantitatively with the calculated change of temperature with the pressure,[3] their value is somewhat doubtful because the system actually observed contained more than one component. It is by no means certain that the liquids used in the dilatometer had no effect on the equilibrium.[4]

The curves OC, O_1C, O_2C meet at an unknown point provided that no other modification of sulfur appears. The position of this point has been calculated from the pitch of the curves OC and O_2C by Roozeboom.[5] Assuming that there is no change in the specific heats of the two solid modifications of sulfur he finds that the inversion temperature should be about 135° and the pressure about 400 atmospheres. At the point C there would coexist, rhombic, monoclinic and liquid sulfur, and at higher pressures the monoclinic sulfur would disappear and the stable monovariant system possible along CD would be rhombic and liquid sulfur. This has not been realized experimentally. As in all diagrams of this kind the divariant systems exist in the fields, we have sulfur vapor bounded by AO_1B, liquid sulfur by AO_1D, rhombic sulfur by DO_1B and monoclinic sulfur by $O CO_1O$. It will be noticed that the monoclinic modification is the only one existing in a closed field and that the overlapping portions of the other fields represent states of labile equilibrium, instable with respect to monoclinic sulfur. The labile

[1] This curve has been followed over a range of 100° for the two modifications of silver iodide. Le Chatelier and Mallard, Comptes rendus, **99**, 157 (1884).
[2] Recueil Trav. Pays-Bas, **2**, 262, 269 (1884).
[3] Ibid. 266.
[4] Cf. Bancroft, Jour. Phys. Chem. **1**, No. 3 (1896).
[5] Recueil Trav. Pays-Bas. **6**, 314 (1887).

states of equilibrium in the case of sulfur show the same characteristics as the corresponding states in the case of water. Addition of a slight quantity of the phase with respect to which the system is instable produces a rapid transformation; but with sulfur there is a case not analogous to any occurring with water. When the change is from one solid phase to the other it does not take place instantaneously. Reicher[1] found that the reaction velocity for the change of monoclinic into rhombic sulfur increased as the temperature fell below 95°, reaching a maximum at about 35° and then decreasing. Ruys[2] took advantage of a winter spent upon the sea of Kara to observe the reaction velocity at a temperature of $-35°$ and found that under those circumstances the time necessary for the change of a given weight of monoclinic into rhombic sulfur was about five hundred times as long as at ordinary temperatures. The reaction velocity is a function of the pressure difference and the absolute temperature. Below the inversion point these work against each other and the point of maximum velocity is the temperature at which the latter effect just overbalances the former. Above the inversion point the two causes work together and there is no decreasing velocity after exceeding a given temperature.

Since the monoclinic is the less stable form at low temperatures, it follows from the Theorem of Le Chatelier that the change to the rhombic modification must be accompanied by an evolution of heat, and this is the case, experimentally.[3] It is impossible to take into account the additions to the diagram due to the existence of other modifications of sulfur because only the rhombic and monoclinic forms have been studied with any approach to thoroughness.[4]

It is not necessary that each allotropic form must be stable at some temperature and pressure, and a very striking instance of this is found in the case of phosphorus.[5] Fig. 3 is the pressure-temperature diagram for this substance. The drawing is not to scale.

---

[1] Recueil Trav. Pays-Bas, **2**, 251 (1883).

[2] Ibid **3**, 1 (1884).

[3] Mitscherlich, Pogg. Ann. **88**, 328 (1853).

[4] Cf. Dammer, Handbuch I, 597-605; Meslans, États allotropiques des corps simple, 34; D. Berthelot, Allotropie des corps simple, 18.

[5] Bakhuis Roozeboom, Recueil Trav. Pays-Bas, **6**, 272 (1887); Riecke's treatment is very inaccurate. Zeit. phys. Chem. **6**, 411 (1890); **7**, 115 (1891).

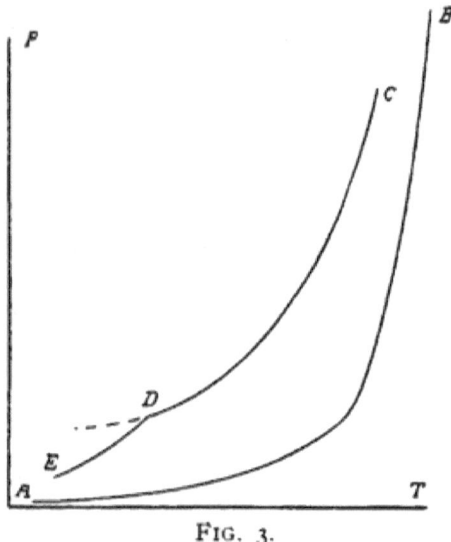

Fig. 3.

AB is the boundary curve for red phosphorus and vapor, ED for yellow phosphorus and vapor and DC for liquid phosphorus and vapor. The curve CDE lies above the curve BA and represents states of labile equilibrium. The only stable forms of phosphorus which have been observed are red phosphorus and phosphorus vapor. The yellow modification and the melted phosphorus are both labile forms, instable in respect to red phosphorus.[1] Whether the curves DC and AB meet at some higher temperature is unknown. If they do meet there can coexist at that point red phosphorus, liquid and vapor, and, from there on, liquid phosphorus can exist in stable equilibrium with the vapor. We get here an interesting application of the statement of Reicher[2] that the reaction velocity is less at temperatures far below the inversion point than in its immediate vicinity. Above 520° the liquid phosphorus changes into the

---

[1] It is probable that the substances classified as "monotropic" by Lehmann (Molekularphysik I, 193-219,) are analogous to phosphorus; one of the solid modifications is always instable with respect to the other without changing into the stable form very rapidly.

[2] Recueil Trav. Pays-Bas. **6,** 251 (1883); Cf. Gernez, Comptes rendus, **100,** 1382 (1885).

red modification so rapidly that only the curve AB can be measured beyond this point. Below 520° it is possible to determine the values of CD, and at ordinary temperatures both the liquid and the yellow phosphorus are fairly permanent even in the presence of the red variety. There is no difficulty in determining the temperature of the inversion point D, 44°, although the nonvariant system, solid, liquid and vapor, is not in a state of stable equilibrium. In Table VI. are the numerical data for phosphorus.[1] Under A are the vapor pressures of ordinary, liquid phosphorus; under B those for red phosphorus. The values in the first pressure-column are expressed in millimeters of mercury; in the other two in atmospheres.

TABLE VI.

Temperature	A	Temperature	A	B
165°	120 mm.	360°	3.2 Atm.	0.6 Atm.
170	173	440	7.5	1.75
180	204	487		6.8
200	266	494	18.0	
209	339	503	21.9	
219	359	510		10.8
226	393	511	26.2	
230	514	531		16.0
290	760	550		31.0
		577		56.0

The relative stability of labile modifications at temperatures far enough below the inversion temperature is further illustrated by the behavior of arragonite and calcite, two modifications of calcium carbonate. On heating, the former changes into the latter; but at ordinary temperatures arragonite is apparently stable even in contact with calcite.[2] The apparent stability of the three modifications of carbon, and of titanic acid,[3] is doubtless due to the very high inversion temperatures in these cases. It is sometimes thought that the occurrence of allotropic modifications is something unusual;

---

[1] Schrötter, Pogg. Ann. **81**, 276 (1850); **Troost and Hautefeuille, Comptes rendus, 76,** 219 (1873).

[2] Rose, Pogg. Ann. **42**, 360 (1837).

[3] Meyerhoffer, Die Phasenregel, 18; Lehmann, **Molekularphysik I, 217.**

but it would probably be quite as near the truth to say that most solids can exist in more than one form,[1] though no special stress has been laid upon this branch of the subject and our knowledge of the possible modifications of many compounds is very rudimentary. In Table VII. are the inversion temperatures of a few of the substances that have already been studied.[2]

The coexistent phases in all these cases are the two solid modifications and the vapor.

TABLE VII.

Mercuric iodide	128.°
Silver iodide	146.
Potassium nitrate	129.5
Ammonium nitrate I	32.4
Ammonium nitrate II	82.7
Ammonium nitrate III	125.5
Silver nitrate	159.5
Lead nitrate I	161.4
Lead nitrate II	219.0
Boracite	265.2
Carbon hexachloride I	44.
Carbon hexachloride II	71.1
Carbon tetrabromide	46.1

[1] Cf. Lehmann, Molekularphysik I, 153-219; D. Berthelot, Allotropie des corps simples; Meslans, États allotropiques des corps simples.

[2] Schwarz, Prize Dissertation, Göttingen, (1892).

TWO COMPONENTS

CHAPTER IV

ANHYDROUS SALT AND WATER

A phase consisting of two or more components is called a compound if it is described by the Theorem of Definite and Multiple Proportions, a solution if this is not the case. A solution may also be defined as a phase in which the relative quantities can vary continuously within certain limits or as a phase of continuously varying concentration.[1] This definition does not confine solutions to the liquid phase, but includes mixtures of gases and of solids. It is a question whether mixtures of gases should be included. In so far as they conform to the Theorem of Dalton they are only mixtures and the variations from that Theorem are scarcely sufficient to permit one to classify them as solutions.[2] Since gases can not form a surface distinct from that of the vessel in which they are contained, they are therefore consolute,[3] *i. e.* miscible in all proportions.

The conception of solid solutions is due to van 't Hoff.[4] Instances of solid solutions are to be found in such salts as potash and ammonia alum, beryllium sulfate and seleniate, ammonium and ferric chlorides, potassium and thallium chlorates and many others.[5] The distinction between isomorphic solutions and mix crystals does not seem necessary, the only difference being that in the mix crystals the single components are not isomorphous. Further examples are the optically homogeneous colored minerals with colorless ground, the glasses and some alloys. Under the same head are probably to

---

[1] Cf. van 't Hoff, Zeit. phys. Chem. **5,** 323 (1890); Ostwald, Lehrbuch I, 606; Nernst, Theor. Chem. 87; Le Chatelier, Équilibres chimiques, 133.

[2] Galitzine, Wied. Ann. **41,** 588, 770 (1890).

[3] Bancroft, Phys. Rev. **3,** 21 (1895).

[4] Zeit. phys. Chem. **5,** 322 (1890).

[5] Roozeboom, Ibid. **8,** 530 (1891); **10,** 148 (1892); Fock, Ibid. **12,** 657 (1893); Stortenbeker, Ibid. **16,** 250 (1895).

be included the cases of occlusion[1] by precipitation so common in analytical chemistry and possibly the gelatinous colloidal hydrates. The absorption of hydrogen and other gases by metals is certainly due in part to the formation of solid solutions as is also the absorption of oxygen and carbonic acid by glass at a temperature of 200° under 200 Atm. pressure.[2] These mixtures show some of the properties of liquid solutions though in a much less marked manner, owing to the resistance to change due to the solid state. Violle[3] found that when a porcelain crucible was heated in charcoal, the carbon diffused through it. Carbon diffuses in perceptible quantities into an iron bar during one day's heating at 250°.[4] Copper has been known to diffuse into platinum and into zinc.[5] Warburg found that it was possible to electrolyze glass between electrodes of sodium amalgam.[6] There occurred an interesting example of the limitations introduced by the solid state. It was possible to pass sodium through sodium glass without any change being visible. If electrodes of lithium amalgam were used, the sodium was replaced by lithium without difficulty; but the glass became opaque and crumbly because this change is accompanied by contraction. On the other hand, since the replacement of sodium by potassium involves expansion it was found impossible to electrolyze a sodium glass between electrodes of potassium amalgam while there was no difficulty when a potash glass was substituted. In the electrolysis of rock crystal further complications were introduced with the crystalline form, it being possible for the current to pass in a direction parallel to the main axis and impossible in the direction perpendicular to it.[7]

In a solution containing two components only, one is called the solvent, the other the solute or dissolved substance.[8] In cases of

---

[1] Schneider, Ibid. **10**, 425 (1892).

[2] Hannay, Chem. News. **44**, 3 (1881).

[3] Comptes rendus, **94**, 28 (1882); also Marsden, Proc. Edinburgh Soc. **10**, 712 (1880).

[4] Colson, Comptes rendus, **93**, 1074 (1881).

[5] Cf. also Roberts-Austen, Phil. Mag. (5) **41**, 526 (1896).

[6] Wied. Ann. **21**, 622 (1884); Tegetmeier, Ibid. **41**, 18 (1890).

[7] Warburg and Tegetmeier, Wied. Ann. **35**, 455 (1888); Tegetmeier, Ibid. **41**, 18 (1890).

[8] Bancroft, Proc. Am. Acad. **30**, 324 (1894); Cf. Story-Maskleyne, Introduction to Fock's Chemical Crystallography.

limited miscibility there is no difficulty in telling which component is solvent and which solute; but when the two substances are consolute there is at present no sure way of deciding at what concentration the change takes place. As this is essentially a quantitative matter, it does not affect the consideration of equilibrium as described by the Phase Rule, and when one component is present in large excess it is safe to speak of it as the solvent. This distinction between solvent and solute does not apply to mixtures of gases,[1] so far as is yet known, an additional reason for not considering them as solutions.

When there are two components it requires four coexisting phases to constitute a nonvariant system, three and two for a monovariant and a divariant system respectively. Since no two components exhibit all the types of equilibrium, it will be better to consider a series of characteristic pairs, each illustrating some new case of equilibrium at an easily accessible temperature and pressure. It will then be possible to classify the different phenomena so as to gain a view of the whole field. Having studied in detail the effect of changes of external pressure and temperature on a system of one component it will not be necessary to repeat this when there are more components unless there is some new feature introduced thereby. The first case to consider is the one where the two components do not crystallize together, exist each in only one solid modification, and there is only one nonvariant system possible at ordinary temperature, two solid phases, solution and vapor. The equilibrium between potassium chloride and water will serve as a type. The graphical representation of this system with the pressure and temperature as co-ordinates is given approximately in Fig. 4.

The nonvariant system, potassium chloride, ice, solution and vapor, is found to be possible experimentally at one temperature and one pressure only, represented in the pressure-temperature diagram by the point O. Any continued change in the external conditions produces finally the disappearance of one of the phases, the temperature and pressure remaining constant so long as all four are present. Which of the two solid phases disappears first on addition

---

[1] The distinction undoubtedly exists in many cases though it lacks experimental confirmation.

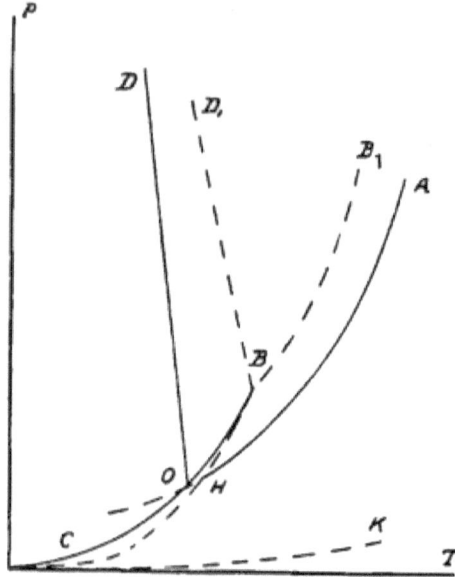

Fig. 4.

of heat depends on the relative quantities of the two, and if present in the same proportion as in the solution they will disappear simultaneously. In this case the whole of the solid will melt or the whole of the solution will freeze without change of temperature, a behavior which is often assumed erroneously to be a criterion of the purity of a compound.[1] The particular mixtures of salts and ice which have a constant melting point were called cryohydrates by Guthrie[2] and were supposed to be compounds. This was shown not to be the case by Pfaundler,[3] Offer[4] and others[5] for the following reasons: The compositions of the supposed compounds did not conform to the Theorem of Definite and Multiple Proportions; the

[1] Cf. Remsen, Organic Chemistry 7.
[2] Phil. Mag. (4) **49,** 1, 206, 266 (1875); (5) **1,** 49, 354, 446; **2,** 211 (1876) **6,** 135 (1878).
[3] Ber. chem. Ges. Berlin, **20,** 2223 (1877).
[4] Sitzungsber. Akad. Wiss. Wien, **81,** II, 1058 (1880).
[5] It is interesting to note that the true explanation was offered by Schultz, Pogg. Ann. **137,** 247 (1869) six years before Guthrie's first paper.

crystals were never transparent and therefore probably not homogeneous; the specific volumes and the heats of solution were additive properties. Another reason for considering the crystals as inhomogeneous mixtures was that alcohol dissolved the ice, leaving the salt; but this point is not well taken, for alcohol will dissolve cupric chloride from the double chlorides of copper and potassium.[1] From the point of view of the Phase Rule it is clear that it is because the two substances do not crystallize together that there are four phases and a constant freezing point. The phenomenon is entirely general and occurs in all cases when a solution, saturated in respect to a solid, is cooled to the temperature at which the solvent begins to freeze out. The cryohydric temperature is the temperature of intersection of a solubility and a fusion curve.[2] The easiest way to prepare a cryohydrate is to cool a saturated solution till the temperature remains constant, pour off the remaining solution and let it solidify, which it will do without change of temperature. In Table VIII. are the cryohydric temperatures of several salt solutions, and

Table VIII

Potassium bromide	$-13.°$	13.9
Potassium chloride	-11.4	16.6
Potassium iodide	$-22.$	8.5
Potassium nitrate	-3.6	44.6
Potassium sulfate	-1.2	114.2
Sodium bromide	$-24.$	8.1
Sodium chloride	$-22.$	10.5
Sodium iodide	$-15.$	5.8
Sodium nitrate	-17.5	8.1
Sodium sulfate	-0.7	165.6
Ammonium bromide	$-17.$	11.1
Ammonium chloride	$-15.$	12.4
Ammonium iodide	-27.5	6.4
Ammonium nitrate	-17.2	5.7
Ammonium sulfate	$-17.$	10.2

[1] Bancroft, Phys. Rev. **3**, 401 (1896); Cf. Ambronn and Le Blanc, Zeit. phys. Chem. **16**, 179 (1895); Küster, Ibid. 525.

[2] Bancroft, Jour. Phys. Chem. **1**, No. 3 (1896). This definition holds only when the solution is saturated in respect to a solid. It is not necessary that what separates should be one of the pure components; it may contain both components as in the case of a hydrated salt.

the compositions of the cryohydric mixtures. The concentrations are expressed in reacting weights of water per reacting weight of salt.[1]

If solid salt be present in excess, the ice will be the first phase to disappear on addition of heat, leaving the monovariant system, salt, solution and vapor. For each temperature there will be a definite pressure in the vapor phase and a definite concentration in the solution at which the system can be in equilibrium and this pressure and this concentration will vary with the temperature. In the diagram, the curve OA represents the pressures and temperatures at which the saturated solution can exist. It is a solubility curve, water being the solvent. Addition of liquid to the solution from outside or from the vapor phase by condensation causes more of the salt to go into solution until the equilibrium concentration is restored ; removal of water, by evaporation for instance, brings about a precipitation of the solute. Both these changes are in accordance with the Theorem of Le Chatelier. Addition of water means a decrease in the concentration of the salt which is neutralized by more salt going into solution. Removal of water increases the concentration and the equilibrium is restored by elimination of the excess of salt. Since it is easier to measure concentrations than vapor pressures it is more familiar to every one that for each temperature there is a single well-defined solubility than that the same is also true for the pressures. In one case it is easy to show the applicability of the Phase Rule to the relation between pressure and temperature. There should be but one temperature at which the vapor pressure of a monovariant system can be equal to the atmospheric pressure, and it is found experimentally that the boiling point of a saturated solution is constant so long as the three phases are present and the barometric pressure remains unaltered.[2] The concentration of the solution changes with the temperature and the direction of this change can be foretold from the Theorem of Le Chatelier. If the solid dissolves with absorption of heat, it will dissolve in greater

---

[1] Guthrie, Phil. Mag. (4) **49**, 269 (1875).
[2] For boiling points of saturated salt solutions, see Landolt and Börnstein's Tabellen, 232.

quantity if the temperature of the system rises, as this change involves addition of heat.¹

Since most salts dissolve in water with absorption of heat, the increasing solubility with rising temperature is just what one would have expected. There are substances known, such as calcium hydrate, sodium, cerium and thorium sulfates,² and calcium isobutyrate which evolve heat on going into solution, and, in all these cases, there is decreasing solubility with increasing temperature.³ As it is not necessary that the heat of solution should have the same sign at all temperatures, it is possible for the solubility of a salt to increase with rising temperature and then decrease as the temperature rises still higher or vice-versa. Examples of the first type are calcium sulfate which reaches a maximum solubility between thirty and forty degrees,⁴ and calcium isobutyrate which has a maximum solubility in the neighborhood of 80°.⁵ Whether the decreasing solubilities of the many sulfates, sulfites, oxalates and carbonates at high temperatures are further illustrations of this is not certain, as it has not been shown that the same substance crystallizes from the solutions at the different temperatures.⁶ An example of the second class where the solubility decreases at first to increase later is to be found in calcium butyrate and possibly in sodium sulfate. This last is said to have a minimum solubility at about 125°.⁷ In all these cases the heat of solution is zero at the temperature of the maximum or minimum solubility, whichever it happens to be. There is no instance known where the sign of the heat effect is not in accordance with the Theorem of Le Chatelier though the contrary has been maintained owing to a false application of the theorem.⁸ The theorem predicts the direction of the change when the system passes from one state of equilibrium to another owing to a change in one or more of the factors of equilibrium. In the particular case in hand, the change

---

¹ Le Chatelier, Équilibres chimiques, 50.
² I am told that this is characteristic of the sulfates of all the rare earths.
³ Cf. Roozeboom, Recueil Trav. Pays-Bas, **8**, 137 (1889).
⁴ Berthelot, Mécanique chimique I, 131.
⁵ Le Chatelier, Comptes rendus, **104**, 679 (1887).
⁶ Étard, Ibid. **106**, 206, 740 (1888).
⁷ Tilden and Shenstone, Phil. Trans. **175**, 23 (1884).
⁸ Chancel and Parmentier, Comptes rendus, **104**, 474 (1887).

of solubility with the temperature can be foretold from the heat evolved or absorbed when we pass from one saturated solution to another. To put it differently, the important point is the sign of the heat effect when the solute is added to an almost saturated solution. It is not proper to consider the heat evolved or absorbed when the solute is added to pure water. This last is the heat of solution usually determined in thermochemistry because it is easier to measure. The heat of precipitation, on the other hand, is very nearly the heat referred to in the Theorem of Le Chatelier. While the heat of solution in the thermochemical use of the term or the heat effect when the solute is dissolved in much water has usually the same sign as the negative heat of precipitation this is by no means always the case. Calcium isobutyrate, at temperatures below 80°, dissolves in a great deal of water with evolution of heat; in a little water with absorption of heat. Reicher and van Deventer showed that the same thing took place with cupric chloride, there being an evolution of heat when the salt dissolved in a large excess of water, and also an evolution of heat when the salt was precipitated from a supersaturated solution.[1] It follows from this behavior that there must be some quantity of water in which one gram of salt will dissolve without either evolution or absorption of heat. This conclusion, which has no theoretical importance, was verified experimentally. The same phenomenon has been observed with the hydrates of ferric chloride.[2] When the heat of precipitation of a solute is zero the solubility does not change with the temperature. This is very nearly realized in the case of sodium chloride. The absolute solubilities of different solids in liquid solvents is a subject about which it is impossible to make any predictions in the present state of our knowledge. There are all degrees of miscibility from barium sulfate which is soluble approximately one part in four hundred thousand of water to pyrogallol which is miscible in nearly all proportions with water. Save for a few empirical generalizations our ignorance is complete.[3]

[1] Zeit. phys. Chem. **5**, 559 (1890).
[2] Roozeboom, Ibid. **10**, 501 (1892).
[3] Cf. Ostwald, Lehrbuch I, 1066; Carnelley, Jour. Chem. Soc. **53**, 782 (1888); Étard, Comptes rendus, **98**, 1276 (1884); Vaubel, Jour. prakt. Chem. (2) **52**, 72 (1895).

The vapor pressure of the monovariant system, salt, solution and vapor is always less than that of the pure solvent at the same temperature provided, as in this case, the vapor pressure of the solute can be disregarded. If the solute has a perceptible vapor pressure of its own, the vapor pressure of the system may be greater or less than that of either component when pure; but the partial pressure of the solvent in the vapor space above the liquid is always less than its pressure in the pure state. That this must be so can be seen by an application of the Theorem of Le Chatelier.[1] Suppose we have a liquid in equilibrium with its own vapor and add a small quantity of some substance soluble in the liquid. There will be a tendency to eliminate the disturbing factor by condensation of vapor, thus reducing its concentration. This change will take place until equilibrium is reached at a diminished vapor pressure for the solvent. This reasoning does not apply to the solute for its concentration might be diminished by increasing its volatility. No cases of this sort have been studied quantitatively as yet, though the theory of distillation with steam can not be worked out completely until this is done. As more of the solute is added, the vapor pressure of the solvent decreases until the maximum concentration or point of saturation of the liquid phase is reached, when the disturbing factor is eliminated by precipitation, the vapor pressure of the system remaining constant. It follows that the boiling point of a solution saturated in respect to a non-volatile solute will always be higher than that of the pure solvent. The vapor pressure of the monovariant system, salt, solution and vapor, increases with rising temperature but not so rapidly as that of the pure solvent, because of the ever greater depression due to increasing solubility. It is conceivable, theoretically, that the lowering of the vapor pressure due to increased solubility might be so great as more than to counterbalance the normal increase conditioned by the heat of vaporization, in which case there would be a decrease of pressure with increasing temperature.[2] This has been realized in the case of calcium chloride.[3] It should be kept in mind that there is nothing in the

---

[1] Bancroft, Jour. Phys. Chem. **I**, No. 3 (1896).
[2] Meyerhoffer (Die Phasenregel, 25,) has made his diagram as if this always occurred.
[3] Roozeboom, Zeit. phys. Chem. **4**, 45 (1889).

Phase Rule to imply that the heat of vaporization of a salt solution is the same as that of the pure solvent. This assumption is involved in the Theorem of v. Babo[1] and in most of the modern quantitative work on vapor pressures, but it is probably not accurate in any case. That it is a very close approximation in many instances is shown by Raoult's work on vapor pressures of dilute solutions.[2] As the temperature rises the vapor pressure of the saturated solution becomes greater up to the critical point of the solution beyond which temperature and pressure, represented in the diagram (Fig. 4) by A, this monovariant system can no longer exist. There are four possible cases each of which seem to have been observed experimentally. The solid may melt and not be consolute with the solution. This occurs with naphthalene and water, sulfur and toluene for instance; the point A is a quadruple point, the nonvariant system being composed of a solid, a vapor and two liquid phases. The solid may melt and mix with the solution,[3] leaving two phases, solution and vapor. An instance of this is silver nitrate and water.[4] The solubility may decrease until the solution and vapor have the same composition and there is left the solid and a phase which is either liquid or vapor as one chooses. This seems to occur in solutions of sulfates in water though the experiments have not been pushed to the critical point.[5] Lastly the solution and vapor may come to have the same composition by increased vaporization of the solute instead of by decreased solubility as in the preceding case. This has been realized by Hannay[6] with potassium iodide and alcohol. He did not work with a saturated solution, but an increase in concentration would not change the character of the phenomenon, only the temperature at which it takes place. In this last case, the critical temperature of the solution lies between the

---

[1] Ostwald, Lehrbuch I, 706.

[2] Comptes rendus, **103**, 1125 (1886); Cf. also, Emden, Wied. Ann. **31**, 145 (1887); Dieterici, Ibid. **42**, 528 (1891).

[3] The curve at this point is a fusion and no longer a solubility curve. This will be considered in detail later.

[4] Étard, Comptes rendus, **108**, 176 (1889).

[5] Étard, Ibid. **106**, 206, 740 (1888).

[6] Proc. Roy. Soc. **30**, 178 (1880).

critical temperatures of the pure components.[1] In the other direction the solubility curve has been followed beyond the cryohydric temperature,[2] but the system is then in a state of labile equilibrium and the addition of the merest fragment of an ice crystal produces a change to a state of stable equilibrium, the temperature rising to the inversion temperature.

Starting from the nonvariant system, salt, ice, solution and vapor, we can pass to the monovariant system, ice, solution and vapor, by supplying heat, provided ice is present in excess. Since the curve OB, representing this series of equilibria, ends at the melting point of pure ice, it is called a fusion curve. The distinction between a fusion and a solubility curve is that the former ends at the melting point of one of the pure components while the latter does not. For each concentration or for each pressure, which is another way of saying the same thing, there is but one temperature at which ice can be in equilibrium with solution and vapor. It is usually assumed that these three phases will be in equilibrium when the vapor pressure of the solid solvent is equal to the partial pressure of that component in the system, solution and vapor,[3] but this is an assumption which is probably never accurately true. It is very doubtful in my mind whether a solution of alcohol in water is in equilibrium with ice at the temperature at which the vapor pressure of the ice equals the partial pressure of the water vapor above the solution. This difference of pressure is probably very slight, in most cases, and its non-existence has seemed so self-evident as to require no proof. The vapor pressures of the fusion curve exceed those of the solid solvent by the partial pressure of the solute and the correction term just referred to. For a non-volatile solute, a solid in the pure state at the temperature of the experiment, both of these variations may be neglected in the present state of our knowledge and the two curves are identical. Under these circumstances the fusion curve OB has a greater vapor pressure for a given temperature than the solubility curve OA, because the solution is more dilute in the first case. Since the solute lowers the vapor pressure

---

[1] Winkelmann, Handbuch der **Physik II, 2,** 668.
[2] Guthrie, Phil. Mag. (4) **49,** 214 (**1875**).
[3] Guldberg, Comptes rendus, **70,** 1349 (1870).

of the solution, it lowers the temperature at which the solution is in equilibrium with the solid solvent. This is best seen from the diagram (Fig. 5).

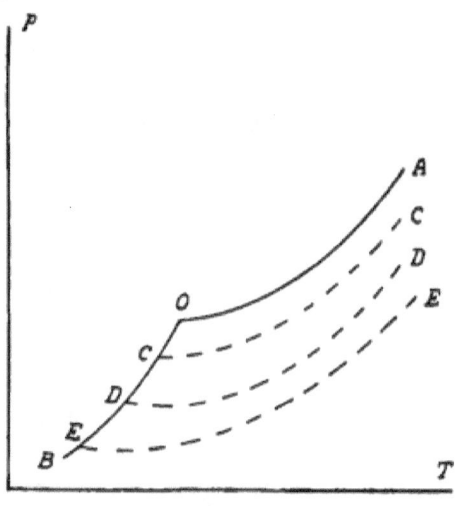

FIG. 5.

OA is the pressure-temperature curve for the liquid solvent in presence of vapor, OB the corresponding curve for the solid solvent and O the freezing point. CC, DD, EE. are the pressure temperature curves for solutions of ever greater concentration. These solutions freeze at the temperatures at which their vapor pressure curves cut the curve OB, namely at C, D and E. The presence of a dissolved substance lowers the freezing point of the solution in all cases where pure solid solvent separates out. Later we shall see that when this last condition is not fulfilled, the freezing point is not necessarily lowered.

The equilibrium between ice, potassium chloride and vapor is represented in Fig. 4 by the curve OC. Any change in external pressure produces a change in the quantity of one of the phases and for each temperature there is but one pressure at which the system can exist. The pressure of this system will be equal to the sum of the vapor pressures of the two solid components if each vapor may be considered as an indifferent gas,[1] an assumption which is never

[1] Nernst, Theor. Chem. 376.

absolutely accurate.[1] As the diagram shows, this system can be in equilibrium only at temperatures below that of the cryohydric point. If a salt is mixed with ice at a higher temperature, the ice will melt and the salt dissolve until one or the other disappears, forming one of the two stable monovariant systems, salt, solution and vapor, or ice, solution and vapor, as the salt or ice is in excess. The ice in melting absorbs heat, the salt in dissolving may evolve or absorb heat. If the latter, as is usually the case, both these changes work in the same direction and the total absorption of heat is very considerable. It is this phenomenon which is utilized in many artificial freezing mixtures. The system, salt, ice and vapor, not being in equilibrium at the temperature of the experiment tends to pass into a state of stable equilibrium. The temperature falls until one or the other solid phase disappears or until the temperature of the cryohydrate is reached. This is the lowest temperature which can be attained at atmospheric pressure with a given freezing mixture, because at and below this temperature the two solid phases can be in equilibrium with each other and will not react. It would be possible by keeping the mixture in an air current to obtain slightly lower temperatures, but the cooling in this case would be due to the heat absorbed in the evaporation of the ice and would have nothing to do with the nature of the other phase. The temperature reached in laboratory experiments is by no means always that of the cryohydrate. The two components are mixed, let us say, at zero degrees and the temperature at once falls. At the same time there is an ever-increasing amount of solution formed which has to be cooled down as well as the rest of the system. The result is that there is reached a point where the heat absorbed by the amount of solution formed in the unit of time is just sufficient to keep the mass of solution already present at constant temperature. What this temperature will be depends on the initial temperature, on the rate of radiation, on the quantity of salt and ice used and on the thoroughness with which the ice and salt have been mixed. Since the reaction velocity is proportional, among other things, to the surfaces of the solid phases in contact, less heat will be absorbed in the unit of time and therefore equilibrium will be reached at a higher temperature

---

[1] Bancroft, Phys. Rev. **3**, 414 (1896).

when the ice is in large pieces than when the two components are ground very fine and intimately mixed. For this reason snow is better than ice, being more finely divided. The most effective arrangement for a salt-ice freezing mixture is to have the mixture in a perforated vessel so that the solution can run off as fast as formed. When the temperature has fallen to the cryohydric temperature the mixture can be placed in an impermeable vessel and used for freezing purposes. It is also economical to take the two components in the proportion in which they occur in the solution at the cryohydric temperature, as the melting point of the cooled solids will remain constant.[1] It is clear that the requisites for a successful freezing mixture are that the cryohydric temperature should be very low and that the heat absorbed per gram of solution formed should be as great as possible. The first criterion is satisfied if the salt is very soluble, the second if the increase of the solubility with the temperture is great. For this reason the freezing mixture, made of common salt and ice, so often used in the laboratory, is not a good one, except on the ground that sodium chloride is cheap. As the heat of solution of this salt is very slight the heat absorbed is little more than the heat of fusion of ice alone. A much better mixture is ammonium nitrate and ice and even better, crystallized calcium chloride. The anhydrous salt can not be used as it evolves heat in taking up six units of water. It will be shown later that this is characteristic of all hydrates. While the freezing mixtures in use are often composed of ice and some salt, this is not essential. Any system which is in instable equilibrium and which absorbs heat in passing into the stable form can be used as a freezing mixture, such as alcohol or sulfuric acid and snow, solid carbonic acid and ether, liquid air at atmospheric pressure.

In all cases where two substances can form a solution and can crystallize from that solution in the pure state, the temperature at which the two components can be in equilibrium with the solution is lower than the fusion point of either of the pure components. This is a necessary consequence of the Phase Rule and the Theorem of Le Chatelier, but it has been entirely overlooked by Étard in some

---

[1] Data for many salts with ice can be found in the papers of Guthrie already cited. Cf. also Landolt and Börnstein's Tabellen, p. 315.

theoretical views advanced by him.¹ He **observed that many substances were very slightly soluble in water. Taking this with the fact that most solubilities decrease with falling temperature, he drew the conclusion that the solubility of any substance at the freezing point of the solvent is zero if the solvent freeze at a sufficiently low temperature. This is wrong. The temperature at which the non-variant system, both solids, solution and vapor, exists is the limiting point for the stable monovariant system, solid solute, solution and vapor, and while it may in some cases approach infinitely near to the fusion point of the solvent it can never reach it. Arctowski²
has published some measurements showing the inaccuracy of Etard's hypothesis but seems to have equally erroneous ideas in regard to what actually happens.**

The fourth monovariant system, **represented by the curve OD, is the one in which there is equilibrium between salt, ice and solution. The direction of this curve is determined by the difference between the sum of the volumes of the solid solvent and solute and the volume of the same masses as solution. It is also dependent on the sign of the heat effect when the two solid phases pass into solution. If the solution is formed with expansion of volume and absorption of heat or with contraction of volume and evolution of heat, increase of pressure will raise the temperature at which the two solid phases can be in equilibrium with the solution. If the solution is formed with expansion of volume and evolution of heat, or contraction of volume and absorption of heat, the curve OD will slant to the left as it is drawn in the diagram instead of to the right, showing that increase of pressure lowers the freezing point of the solution. The only difference between this case and that where there is only one component is that the change from pure solid to pure liquid is always accompanied by an absorption of heat, while it is conceivable that two substances might exist such that the heat of solution evolved would be greater than the heat of fusion of the other substance. No instance of this is on record to the best of my knowledge. The Theorem of Le Chatelier enables us to predict that in such a case the solid mass would liquefy on cooling.**

[1] Ann. chim. phys. (7) **2**, 268 (1894).
[2] Comptes rendus, **121**, 123 (1895).

With water as solvent, the most usual case is that the volume of the solution is less than that of the two solid components, while the reverse is true for most saturated solutions in other solvents. Little is known experimentally of the course of this curve,[1] but it seems probable that it will either approach asymptotically the pressure of the system, solid and liquid solvent, or that it will terminate at the intersection with the curve for solid solute, solid solvent and solution provided the vapor pressure of the solute may be neglected.

The different divariant systems exist in the fields limited by the boundary curves and by the curves for the pure components. The system, unsaturated solution and vapor, can exist at a series of temperatures and for each temperature at a series of pressures depending on the concentrations. If the temperature and concentration are both fixed, the pressure is determined thereby. If the solution is heated in an open vessel so that the solvent can distill off, the boiling point will rise with increasing concentration until the solution becomes saturated ; after which it remains constant. The vapor pressure of a dilute solution is less than that of the pure solvent if the partial pressure of the solute may be neglected.[2] For all practical purposes, potassium chloride comes under this head and its vapor pressure may be considered equal to zero.[3] I shall discuss the limiting pressure and temperature for the possible divariant systems on this assumption, showing afterwards the changes necessitated by dropping it. For purposes of reference I have added to the diagram (Fig. 4) the curves BB_1 and BD_1 which represent the equilibrium between water and vapor, water and ice respectively. The curve for ice and vapor coincides with the curves BO and OC, so far as we know now, when we neglect the vapor pressure of the potassium chloride. All possible temperatures and pressures at which there can be equilibrium between solution and vapor lie within the space $AOBB_1$. The points B_1 and A, the critical points of the solvent and solution respectively, will coincide only if the solubility of the solute

---

[1] Roloff, Zeit. phys. Chem. **17**, 348 (1895).

[2] If we eliminate the natural and the induced vapor pressure of the solute as well as the effect due to surface tension, BOC becomes a continuous curve with no change of direction at O.

[3] This is not strictly true. Cf. Bailey, Jour. Chem Soc. **65**, 445 (1894).

becomes zero at the critical temperature and pressure for the solvent. It has already been pointed out that this seems to occur very nearly for many sulfates, sulfites, carbonates and oxalates[1]; but this is not a general phenomenon, and even in these instances the conclusion is based on extrapolation in an empirical formula and not on experimental data. In the field bounded by the line AOC and the temperature axis there is equilibrium between salt and water vapor; and here, too, the equilibrium is not settled definitely till two variables are determined arbitrarily, the pressure and the temperature or one of these and the concentration. Practically, the only change of concentration is in the vapor phase; but, theoretically, the density of the solid phase, and therefore its volume concentration, must change with changing pressure or temperature. If the pressure is increased above that of the saturated solution, represented by the curve OA, vapor will condense, forming a saturated solution. In other words, an anhydrous salt is deliquescent when the pressure of water vapor in the atmosphere is greater than the vapor pressure of the saturated solution and is permanent when it is less. If instead of diminishing the pressure upon a saturated solution reaching the divariant system, solid solute and vapor, we increase it there is formed the system, solution and vapor, if there is only a small amount of undissolved solid and a large amount of vapor, or the system, salt and solution, if there is an excess of the solid phase. This new divariant system, existing only in the field AOD, can have different concentrations and pressures at the same temperature; but there is a definite solubility for each pressure at each temperature. The change in the concentration of the solution is in the direction predicted by the Theorem of Le Chatelier. If the volume of the solid plus the volume of the solvent is greater than the volume of the resulting solution, increasing pressure means increasing solubility, otherwise it involves decreasing solubility, the system being always kept at constant temperature.[2] With most salts there

---

[1] Étard, Comptes rendus, **106**, 206, 740 (1888); Ann. chim. phys. (7) **2**, 546 (1894).

[2] Braun, Wied. Ann. **30**, 250 (1887); v. Stackelberg, Zeit. phys. Chem. **20**, 337 (1896); Ostwald, Lehrbuch I, 1044-1047. It is incorrect to say that the change in the solubility depends also on the sign of the heat of solution. In

is a decrease of volume when dissolving in water which is so great in the case of copper sulfate that a dilute solution of that salt occupies less volume than the pure solvent alone.[1] There are cases known where there is an increase of volume when the salt goes into solution, the most notable instance being ammonium chloride. In this case, increase of pressure involves precipitation of the salt.

In the field $DBOD_1$ there can exist the system, salt and solution, as we have just seen and, under proper conditions, the system, ice and solution. This divariant system has been very little studied, the only interesting investigation of the subject being a paper by Colson.[2] He showed that if the concentration and temperature be fixed the pressure has a definite value which is, of course, a necessary consequence of the Phase Rule. In the field DOC there is equilibrium between salt and ice. The divariant system, ice and vapor, can not be realized if the vapor pressure of potassium chloride be treated as equal to zero. If this assumption be given up, the curve for the equilibrium between ice and its own vapor will no longer coincide with BOC, but will be represented by the line BHC. At the same time it will be necessary to add to the diagram the dotted line CK, showing the vapor pressure of the pure solute at different temperatures. Ice, water vapor and the vapor of the solute can exist in the closed field, COBHC; the field for salt and vapor can not extend below CK; the field for ice and solution is now bounded by D_1BHOD, the other fields not being changed necessarily. While the direction and position of the boundary curves may change a good deal with a volatile solid solute as one component there is but one displacement that calls for particular comment. The boundary curve OA may intersect the curve BB_1 and if this takes place at a pressure of less than one atmosphere we shall have the phenomenon of a solution saturated in respect to a

---

Braun's formula, there appear the heat of solution and the change of solubility with the temperature. As these two always have the same sign, it disappears from the equation. This seems to have been overlooked both by Braun and by Ostwald. Cf. Lehrbuch I, 1046.

[1] Favre and Valson, Comptes rendus, **79,** 936, 1068 (1874); Cf. MacGregor, Zeit. phys. Chem. **9,** 231, 236 (1892).

[2] Comptes rendus, **120,** 991 (1895).

solid boiling at a lower temperature than the pure solvent. I know of no case where this has been observed but this proves nothing. Most solids are very slightly volatile at ordinary temperatures; with the consequence that in a boiling saturated solution the decrease in the partial pressure of the solvent is not compensated by the partial pressure of the solute. The conditions necessary for an example of this type are a solid with a high vapor pressure and a liquid in which the solute shall be sparingly soluble at the boiling point.

Returning to the particular case of potassium chloride and water and the assumption of a non-volatile solute, it is clear from the diagram that if a system composed of salt and vapor is subjected to an external pressure continuously greater than its own and is at the same time kept at constant temperature, there are three cases to be considered; when the temperature is above the freezing point of the pure solvent, when it is between the fusion point of the pure solvent and the cryohydric temperature of the mixture, and when it is below this last temperature. In the first case, there will be compression of the vapor phase without condensation until the vapor pressure of the saturated solution is reached when there will be formed the monovariant system, salt, solution and vapor. The pressure will then remain constant until either the salt or the vapor disappears, depending on the relative quantities of each, leaving the divariant system, solution and vapor, or the one, salt and solution. If the former, further increase of pressure will result in the formation of the trivariant system, solution. This will also be formed eventually by the compression of salt and solution if the salt dissolves with contraction; otherwise the final state will be salt and solution unless the salt should be liquefied by the extreme pressure. This last is not probable as it could occur only in case the salt expanded on melting or that what has been called the solvent dissolved in the salt with expansion of volume. Between the freezing point of the pure solvent and the cryohydric temperature of the system, the result will be the same if the salt is in excess. If this is not so, there will be formation, as before, of solution and vapor; but with increasing pressure, at length, ice will separate and the pressure will then remain constant until the vapor phase has disappeared. The pressure will increase again, the ice dissolving until there is left only solution. If the system is subjected to pressure at temperatures lower than that of the cryohydric point, there will be formed, as first visible change, the monovariant system, salt, ice and vapor; the vapor will condense forming the divariant system, ice and salt, which under further compression, will melt, giving salt, ice and solution. If the salt dissolves with contraction, increase of pressure will cause first one and then the other of the solid phases to disappear, leaving only the solution phase at last. In other words, in all cases where the solvent contracts in fusing and there is decrease of volume and absorption of heat when the solid solute dissolves in the liquid solvent, the final result of compression at constant temperature is the formation of the trivariant system, solution. If these conditions are not fulfilled, the course of events will be somewhat different; but our knowledge of the subject is too limited to permit of making a detailed consideration of all possible cases. If we make the rather plausible assumption [1] that at sufficiently high pressures the change of the boundary curve for solid solvent, solid solute, and solution, wtih the temperature has the same sign as the corresponding value for the boundary curve for solid and liquid solvent, we can say that the final result of compression at constant temperature will be formation of solution if the solvent contracts on fusing and the formation of the two solid phases if the solvent expands when passing from the solid to the liquid state.

If heat be withdrawn from the system, solution and vapor, kept at constant volume, the ordinary case will be a decrease of pressure and temperature until the boundary curve for salt, solution and vapor is reached. The changes of pressure and temperature will be represented by that curve and the curve for the two solid phases and vapor. The assumptions made here are that the vapor phase does not disappear at the cryohydric point either on supplying or withdrawing heat and that the solute is more soluble at high than at low temperatures. The first assumption need not be considered as the remarks about the behavior of a system of one component under similar circumstances apply here with the addition of the solid solute

[1] Cf. however the behavior of sodium chloride and water under pressure. Braun, Wied. Ann. **30**, 262 (1887).

as an extra phase. If the solute disolves with evolution of heat, the monovariant system formed on cooling at constant volume will be the one composed of solid solvent, solution and vapor. That curve will represent the pressures and temperatures of the system until the cryohydric point is reached. The change on cooling further will be the same as in the previous case.

CHAPTER V

HYDRATED SALTS

If we drop the condition that the solvent and solute shall not crystallize together, a number of new solid phases become possible introducing distinct changes in the conditions of equilibrium. It will be best to confine the discussion for the present to the cases in which the two components crystallize in definite and multiple proportions, in other words with formation of compounds and not of solid solutions. When the solvent is water, the compound formed is called a hydrate or a hydrated salt and the water is termed water of crystallization. It is possible also to have benzene, alcohol, ammonia, hydrochloric acid and many other volatile substances crystallizing with a practically non-volatile body in definite, discontinuous amounts. These crystals are often regarded as so-called "molecular" compounds in contradistinction to so-called "chemical" compounds; but this distinction cannot be considered valid until accurate definitions of these hypothetical classes are given. While it is by no means certain or even probable that a radical distinction cannot be drawn between a hydrated salt and calcium carbonate, for instance, none such has been drawn as yet. As an example of the changes in equilibrium introduced by the possibility of the solvent and solute crystallizing together we will consider the case of sodium sulfate which crystallizes from aqueous solutions in the form of Na_2SO_4, of $Na_2SO_4 7H_2O$ and of $Na_2SO_4 10 H_2O$, depending on the conditions of the experiment. The pressure-temperature diagram for sodium sulfate and water is shown in Fig. 6. It is not drawn to scale.

OO_1, OB, OC and OD are the curves representing the equilibria for hydrate, solution and vapor; ice, solution and vapor; hydrate, ice and vapor; hydrate, ice and solution. O is the cryohydric point just as in Fig. 4 and the remarks in regard to the equilibrium between potassium chloride and water apply to these four curves also, there being no change introduced by the presence of a hydrated salt instead of an anhydrous one. If, however, we add heat to the system hydrate, solution and vapor, kept at constant volume, there will be a rise of temperature and pressure, the system passing along the curve OO_1. At O_1 a new solid phase appears in the form of the anhydrous salt and there is present a new nonvariant system composed of hydrate, anhydrous salt, solution and vapor. The point O_1 is therefore an inversion point and must be situated at the intersection of four boundary curves. This is the case and the curves OO_1, AO_1, CO_1, D_1O_1, represent the possible temperatures and pressures for the four monovariant systems, hydrate, solution and vapor; anhydrous salt, solution and vapor; hydrate, anhydrous salt and vapor; hydrate, anhydrous salt and solution.

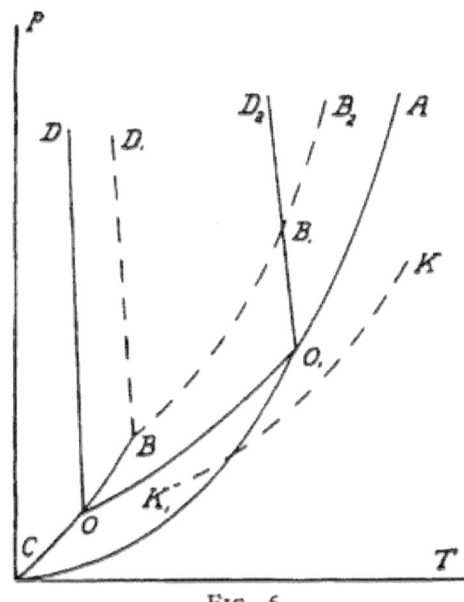

FIG. 6.

The temperature at which the nonvariant system can exist is $32.6°$ and the pressure 30.8 millimeters of mercury.[1] Any continued supply or withdrawal of heat or work results finally in the

[1] Cohen, Zeit. phys. Chem. **14**, 90 (1894).

disappearance of one of the four phases without change of temperature or pressure. The curve AO_1, if prolonged, will be found to lie below the curve OO_1. Since it represents a labile equilibrium at all temperatures below 32.6°, we see that here the less stable system has a lower vapor pressure than the more stable one. Although this is contrary to our previous experiences, it is a perfectly general result. The more stable system is the one with the lesser concentration and therefore the greater vapor pressure. There are two forces acting in opposite directions, the tendency of the vapor to distill from a place of higher to one of lower pressure, and the tendency of the excess of solute to precipitate from the more concentrated solution. Either of these changes would bring about equilibrium alone but the second is the stronger. If the two solutions are placed under a bell-jar we shall get distillation and the more concentrated solution as the stable one ; if the two solid phases are brought into the same solution the stable form is the one which is in equilibrium with the more dilute solution. The important factor in determining the equilibrium is the change in the vapor pressure of the solute. The more concentrated the solution the higher the partial pressure of the solute and stable equilibrium is reached, experimentally, when the more stable solid phase is present, the one with the lower vapor pressure. When the two saturated solutions are connected by the vapor phase only, there is distillation of water in one direction and of solute in the other ; but the velocity of the first reaction being much greater than that of the second the two solutions come to the same concentration before any measurable amount of the solute has been transferred. It is to be noticed that although the partial pressure of the solute may be infinitely small, it exerts the controlling influence in regard to the stability of the system when both solid phases are added. ' It is too often assumed that a value may be neglected in respect to all measurements because it can be neglected in respect to one.

From the existence of these two solubility curves at a given temperature, say 30°, it is clear that it is not sufficient to speak of a saturated solution without defining the solid phase with respect to which it is saturated.[1] As has been said, the solution saturated at 30° with

---

[1] Ostwald, Lehrbuch I, 1036.

respect to anhydrous sodium sulfate contains more of the solute than the solution saturated with respect to $Na_2SO_4 10H_2O$ and is therefore in a state of labile equilibrium, stable only so long as no solid hydrated salt is present. There is yet another saturated solution which can exist at this temperature, the solid phase being $Na_2SO_4 7H_2O$. This system is instable both with respect to the anhydrous salt and to the decahydrate, being more soluble than either. The vapor pressure of the solution is therefore less than that of either of the other solutions and is represented by the dotted line KK_1. This monovariant system differs from the other two in that at no temperature does it represent a state of stable equilibrium. In this it is analogous to the yellow phosphorus which is labile both as solid and as liquid. The crystals of $Na_2SO_4 7H_2O$ can be obtained by addition of alcohol to a mixture of sodium sulfate and water.[1]

Starting from the nonvariant system it is possible to reach the monovariant system, hydrate, anhydrous salt and vapor, by keeping the external pressure constantly less than that of the nonvariant system until the whole of the solution has disappeared. According to the Phase Rule this new system must have a definite, unchanging vapor pressure at each temperature so long as the three phases are present, entirely independent of the absolute mass of any of them. This is found to be the case experimentally. If the external pressure upon the system be increased there will be condensation of water vapor and formation of hydrate at the expense of the anhydrous salt. If it be decreased the hydrate will lose water, efflorescing as it is called. While this increase and decrease of external pressure can be brought about theoretically by considering the system in a closed vessel with a movable piston, such as a barometer tube, and increasing or decreasing the volume of the vapor phase and therefore of the system, it is often convenient, practically, to work at constant volume. This can be done by introducing into the vapor phase some substance, such as pure water which will give off water vapor at a higher pressure than that of the system under consideration or, in the reverse case, by introducing some such substance as strong sulfuric acid which will take up water at a less pressure than that of the system. Another way in which the external pressure may be

---

[1] Dammer, Handbuch II 2, 156.

kept less than the equilibrium pressure of the system without using a closed vessel and a movable piston is by passing a current of dried air or other gas through the vessel in which the system is placed. By this means the water vapor is carried off as fast as formed and there will be no chance for the concentration of water in the vapor phase to rise to the value corresponding to the equilibrium pressure of the system. This last method is of use only as a means of removing a given phase rapidly and cannot be employed for an accurate study of equilibrium because it introduces new components into the system, thereby changing the whole problem.

The vapor pressures of the system, hydrate, anhydrous salt and vapor, cannot be higher than those for the solution saturated in respect to the hydrate, because water would then condense, building the latter system at the expense of the former, which does not happen experimentally. The curve O_1C can not, therefore, lie above the curve O_1O, and often lies a good way below it. In Table IX

TABLE IX

Temp. 20°	Solution	Salt	Temp. 20°	Solution	Salt
$CaCl_26H_2O$	5.4	2.3	$Na_2CO_312H_2O$	16.0	10.1
$SrCl_26H_2O$	11.4	5.6	$Na_2SO_410H_2O$	15.7	13.9
$MnCl_24H_2O$	8.0	3.8	$Na_2SO_47H_2O$	15.0	10.5
$NiCl_26H_2O$	8.0	4.6	$MgSO_47H_2O$	14.5	10.3
$CoCl_26H_2O$	9.0	4.0	$CuSO_45H_2O$*	58.0	30.0
$NaBr4H_2O$	9.6	7.6	$MgCl_26H_2O$	5.7	1.8
$SrBr_26H_2O$	9.1	1.7	$NaI4H_2O$	5.4	1.5

* Vapor pressure at 45° instead of 20°.

are the vapor pressures in millimeters of mercury of certain saturated solutions and of the partly effloresced crystals. As will be noticed the differences between the two columns are quite considerable in many instances. In some cases the two sets of values coincide within the limit of experimental error, as is shown in Table X[1] and still more strikingly in Table XI.[2]

[1] Lescœur, Ann. chim. phys. (6) **19**, 533; **21**, 511 (1890); (7) **2**, 78 (1894).
[2] Joannis, Comptes rendus, **110**, 238 (1890).

Table X

	Solution	Salt	Temp.
$BaBr_2 2H_2O$	10.7	10.6	20°
$BaBr_2 2H_2O$	124.0	124.0	60
$MgBr_2 6H_2O$	3.4	3.3	20
$MgBr_2 6H_2O$	166.0	166.0	100
$CdBr_2 4H_2O$	10.0	9.0	20
$CdBr_2 4H_2O$	122.0	124.0	60
$BaI_2 6H_2O$	8.4	8.4	20
$BaI_2 6H_2O$	58.0	60.0	60
$MnBr_2 4H_2O$	5.0	5.0	20
$MnBr_2 4H_2O$	202.0	200.0	100

Table XI

	Pressure	Temp.
$1gNH_2Na + 1.669gNH_3$	169.7	0°
$1gNH_2Na + 0.460gNH_3$	169.7	0
$0.971gNH_2Na + 0.029gNa$	169.7	0
$0.487gNH_2Na + 0.513gNa$	169.7	0
$0.108gNH_2Na + 0.892gNa$	169.7	0
$0.043gNH_2Na + 0.957gNa$	169.65	0
$1gNH_2Na + 0.46gNH_3$	117.0	−10
$0.7gNH_2Na + 0.30Na$	117.3	−10
$0.39gNH_2Na + 0.61gNa$	117.0	−10
$0.19gNH_2Na + 0.81gNa$	117.1	−10

The data in Table XI are very interesting because they were collected for the express purpose of showing that the vapor pressures of an efflorescing compound could have the same value as the vapor pressures of the corresponding saturated solution for a series of temperatures. Roozeboom[1] has stated that, in his opinion, this could occur only at the inversion temperature, but the facts do not seem to have borne him out in this view. It is clear that these two curves can not actually coincide; but the difference between them may be a difference in the partial pressures and not in the total vapor pressure. In a case of this sort the concentration in the vapor phase of the so-

---

[1] Comptes rendus, **110**, 135 (1890).

called non-volatile component is of great importance in spite of its not having a measurable value. All that the Phase Rule states is that the vapor phase for the one monovariant system can not be identical in every respect with the vapor phase for the other monovariant system. When the curves O_1O and O_1C coincide within the limits of experimental error, it seems probable that the continuation of AO will lie below O_1C. Under these circumstances if a mixture of anhydrous and hydrated salt be placed in a beaker and the instable saturated solution of the anhydrous salt in another beaker, both under the same bell-jar, there will be distillation from the first to the second and spontaneous formation of an instable system at the expense of the stable one. As no instance of this has yet been studied it is hardly worth while to draw any more conclusions till this has been done.

Little is known about the vapor pressures of the system, sodium sulfate heptahydrate, anhydrous sodium sulfate and vapor, though this curve of course lies below KK_1.[1] This curve is not given in the diagram but it exists below $25°$, the point at which the continuation of AO_1 would cut KK_1. Above this temperature we should expect that the crystals of sodium sulfate heptahydrate would change into anhydrous salt, solution and vapor, instead of effloresing. It is not known whether an equilibrium can be established between the two hydrated sodium sulfates and vapor, but this could probably be realized. Usually the decahydrate effloresces to the anhydrous salt without formation of the heptahydrate. Although this question of efflorescence has not been worked out very carefully there is little doubt that a hydrated salt effloresces normally with formation of the solid phase which appears at the next higher stable inversion point.

Since the vapor pressure of the system hydrate, effloresced salt and vapor, increases experimentally with rising temperature, it follows from the Theorem of Le Chatelier that the decomposition of the hydrate into its dissociation products is accompanied by an absorption of heat or that the formation of the hydrate is attended by an evolution of heat. This can be shown experimentally by adding an excess of anhydrous sodium carbonate to water when a distinct rise of temperature will be observed. If water be added to crystal-

---

[1] Cf. Table IX.

lized sodium carbonate, there will be a fall of temperature because the solubility of the hydrate increases with rising temperature and therefore heat is absorbed in the process of dissolving. If the heat evolved in the change from the anhydrous to the hydrated salt be greater than the heat absorbed in passing from the hydrate to the saturated solution, there may be heat evolved in the change from anhydrous salt to saturated solution of the hydrate even though the latter be more soluble in warm than in cold water. This is probably what Ostwald has in mind when he says:[1] " It is a familiar rule that salts which dissolve in water with evolution of heat crystallize with water of crystallization ; anhydrous salts dissolve with absorption of heat." In this form the statement is incorrect.[2] It is also indefinite because Ostwald has not specified whether he refers to the thermodynamical or the thermochemical heat of solution.

The curve O_1D_2 representing the equilibrium between hydrate, anhydrous salt and solution, introduces no new features and needs no discussion. It will be noticed that in the diagram, Fig. 6, the curve KK_1, for the system, $Na_2SO_4 7H_2O$, solution and vapor, cuts at some unknown point the curve O_1C, for the system, $Na_2SO_4 10H_2O$, Na_2SO_4 and vapor. This intersection does not represent a new inversion point because it is not the locus of the curves for four monovariant systems but the intersection of two curves for two monovariant systems which have only one phase in common, the vapor phase. If we could measure the vapor pressure of the sodium sulfate we should find that at the intersecting point the total vapor pressures of the two systems were equal but that the partial pressures were not. There is only an apparent identity in the two vapor phases and therefore no reason for assuming that there might be equilibrium between the two systems

In order to determine the boundaries of the fields in which the divariant systems can exist, I have added the dotted line BB_2 which is the pressure-temperature curve for pure water in equilibrium with its own vapor. Solution and vapor can be in stable equilibrium in the field AO_1OBB_2 ; solution and anhydrous salt in AO_1D_2 ; solution

[1] Lehrbuch II, 800.
[2] Cf. Roozeboom, Recueil Trav. Pays-Bas, **8**, 111 (1889).

and hydrate in D_2O_1OD; anhydrous and hydrated salt in D_2O_1C; hydrated salt and vapor in O_1OCO_1; while anhydrous salt and vapor can exist in the field bounded by AO_1C and the temperature axis. The distribution of the fields round the point O is practically the same as for the system, potassium chloride and water, substituting $Na_2SO_4 \cdot 10H_2O$ for KCl. The only difference is that the divariant system, hydrate and vapor, can not exist at pressures lower than those of the curve O_1C. Since we are considering stable equilibrium only, the hydrate $Na_2SO_4 \cdot 7H_2O$ does not enter into the discussion.

From the diagram we can predict the behavior of hydrates of this type when heated. When the temperature of the point O_1 is reached the hydrate will seem to melt with precipitation of salt, forming the nonvariant system, hydrate, anhydrous salt, solution and vapor, a further addition of heat causing the disappearance of the hydrate. The temperature at which this change takes place in such salts as sodium sulfate and sodium carbonate is not the melting point of the hydrate but the inversion point. Later we shall study hydrates which have true melting points. Hydrate and vapor, being a divariant system, can exist at more than one pressure for a given temperature, and it may be well to specify in words the conditions under which hydrates change when exposed to the air. We have already seen that an anhydrous salt is permanent when the pressure of water vapor in the atmosphere is less than the pressure of the saturated solution and is deliquescent when it exceeds that value. Since the field for hydrate and vapor lies between O_1C and O_1OC it follows that at all temperatures, above that of the cryohydric point O, a hydrated salt will effloresce when the partial pressure of water vapor in the atmosphere falls below the pressure of the system, hydrate, effloresced salt and vapor; will deliquesce when it exceeds the vapor pressure of the saturated solution and will be permanent at intermediate values. Whether the range of pressures over which the hydrate is permanent is an extended one or not depends on the relative positions of the curves O_1O and O_1C at the temperature of the experiment. At temperatures below that of the cryohydric point the hydrate will effloresce under the same conditions as before; but if the pressure of aqueous vapor in the atmosphere exceeds the value of OC for that temperature, there will be a precipitation of ice, the hydrate remaining unchanged.

The general results which have just been obtained are not confined to reactions between salts and water. The same phenomena will recur in all cases in which the solute and solvent form a solid compound, provided that it is not possible to have two co-existing liquid phases, a case which will be treated by itself. It is to be remembered that if each component has a perceptible vapor pressure this will have an effect on the extent of the fields in which the different divariant systems can exist. The phenomenon of a constant vapor pressure at each temperature for the system, hydrate, effloresced salt and vapor, will occur in all cases in which a solid compound dissociates into a solid and a vapor. This has been shown to hold for the compounds formed by the action of ammonia on the silver haloids,[1] on ammonium bromide[2] and on metallic sodium.[3] A more interesting case than any of these is the dissociation of calcium carbonate into calcium oxide and carbonic acid. There is equilibrium between the three phases only at one pressure for each temperature, at the dissociation pressure so-called.[4] The divariant system, calcium carbonate and carbonic acid, on the other hand, can exist at a given temperature under any pressure between the dissociation pressure and the pressure at which carbonic acid condenses to a liquid in the presence of calcium carbonate. Isambert[5] found that at temperatures in the neighborhood of 1100° C, the dissociation pressure of barium carbonate was so low that the salt was practically not decomposable by heat in open vessels owing to the amount of carbonic acid in the air. By passing a current of nitrogen through the apparatus, thus keeping the partial pressure of the carbonic acid down towards zero, the compound effloresced completely with formation of barium oxide. The same effect was obtained by mixing powdered carbon with the salt. The carbon combined with the carbonic acid forming carbon monoxide and keeping the partial pressure of the carbonic acid at a minimum value. It may be urged that all this is not strictly analogous to the behavior of a hydrated

---

[1] Isambert, Comptes rendus, **66**, 1259 (1868); **70**, 456 (1870).
[2] Roozeboom, Recueil Trav. Pays-Bas, **4**, 355 (1885).
[3] Joannis, Comptes rendus, **110**, 238 (1890).
[4] Debray, Ibid., **64**, 605 (1867); Le Chatelier, Ibid. **102**, 1243 (1886).
[5] Ibid., **86**, 332 (1878).

salt because the dissociation pressure of calcium carbonate is practically zero at room temperatures, and at higher temperatures carbonic acid is a gas and not a vapor; but it has not been shown that this fact introduces any fundamental distinction.

Since concentrations are more easily measured than vapor pressures, it is sometimes advantageous to use the concentration and temperature as co-ordinates. The resulting diagram is not so general as the pressure-temperature diagram because it can be applied only to phases of varying concentration; but it conveys certain information that is only obtainable indirectly from the other diagram. Since people have studied, almost exclusively, monovariant systems composed of a practically non-volatile solute and a volatile solvent,

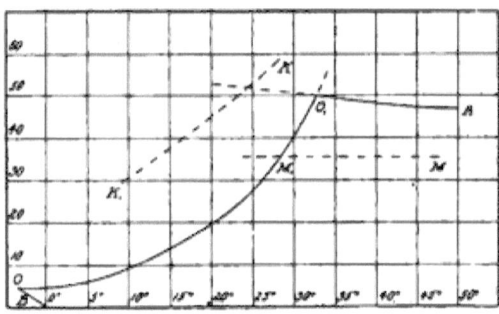

FIG. 7.

the concentration of the liquid phase is the only one that is ordinarily tabulated. This is not necessary and is due to our very complete ignorance of the composition of vapor phases containing two components. In Fig. 7 is shown the concentration-temperature diagram for sodium sulfate and water. The ordinates are grams of sodium sulfate in one hundred grams of water and the drawing is to scale; but the distance equal to one degree is ten times as great for the region—1° to 0° as for the rest of the diagram. The system is supposed to be under its own pressure.

The curve BO is the fusion curve showing the concentrations and temperatures at which the solution can be in equilibrium with ice. This curve runs to the left because increasing concentration causes increasing depression of the freezing point. At the cryohydric point O, the hydrate $Na_2SO_4 \cdot 10H_2O$ begins to crystallize out and we have the first inversion point, the temperature being $-0.7°$ and the concentration being about 4.8 grams per hundred of water. With increasing concentration the system passes along the curve OO_1, the solubility curve for $Na_2SO_4 \cdot 10H_2O$. At O_1 anhydrous sodium sulfate crystallizes from the solution forming the second nonvariant system, the temperature being 32.6° and the concentration about 49.8 grams. Beyond this temperature the concentration does not increase because anhydrous sodium sulfate dissolves with evolution of heat. The curve O_1A is the concentration-temperature curve for the solution saturated in respect to anhydrous sodium sulfate. The line OO_1A is therefore not a continuous curve but two curves meeting at an angle. This is a characteristic phenomenon and one may say that in all cases where a solubility curve shows a "break" or discontinuous change of direction some new substance is separating from the solution. Conversely, if the solubility curve has a continuous change of direction the solution is saturated in regard to the same substance: but this does not apply to the intersection of a fusion and solubility curve, a case which will receive special consideration. The curve OO_1 has been realized only with great difficulty beyond the inversion point; but the curve AO_1 has been followed some fifteen degrees below 32.6°. The solution, thus obtained, contains more of the solute than the solution saturated at the same temperature in respect to $Na_2SO_4 \cdot 10H_2O$ and is therefore instable in the presence of a crystal of the hydrate. The curve KK_1 shows the concentrations at different temperatures of solutions saturated in respect to $Na_2SO_4 \cdot 7H_2O$. As this curve lies above the broken line OO_1A the equilibrium is always labile. At the intersection of AO_1 with KK_1 in the neighborhood of 25° there is possible the labile nonvariant system, Na_2SO_4, $Na_2SO_4 \cdot 7H_2O$ solution and vapor, instable in respect to $Na_2SO_4 \cdot 10H_2O$. I am not aware that this has been observed experimentally.

In the field bounded by AO_1OB and the temperature axis there is stable equilibrium between unsaturated solution and vapor. If a solution having the concentration and temperature represented by the point M be cooled in a closed vessel, the changing state of the system will be represented by the horizontal dotted line MM_1. At

M_1 the solution has the concentration of the saturated solution and on further cooling the solid phase should appear, the system passing then along the line M_1O. If the solution be not jarred, it often happens that no precipitate is formed and the state of the system is represented by some point on the prolongation of MM_1. As the solution contains more dissolved substance than the saturated solution it is instable with respect to the solid phase and is said to be supersaturated.[1] In many cases shaking will cause a precipitation of the excess; but the only certain method of causing crystallization is the addition of a crystal of the substance in respect to which the solution is supersaturated or a crystal of an isomorphous compound[2] The phenomenon of supersaturation is not confined to salts in water, being a more or less developed characteristic of all solutions. Substances differ very greatly in the ease with which they form supersaturated solutions and very little is known of the cause of this behavior. In aqueous solutions it may be said, as a rule, that the salts which separate with water of crystallization form supersaturated solutions more readily than salts which separate in the anhydrous state. Sodium sulfate, sodium carbonate and the alums are good instances of the first type, potassium nitrate and sodium chloride of the second. This is not a general statement because silver nitrate and sodium chlorate, for instance, crystallize in the anhydrous form and yet form supersaturated solutions with great readiness, to say nothing of the fact that most obstinate cases occur with organic solutes in organic solvents. Ostwald[3] has pointed out that there is an evident connection between the power to form large crystals and the tendency to form supersaturated solutions. A salt which forms highly supersaturated solutions is one which tends to precipitate only on a crystal already present while a salt which forms supersaturated solutions with difficulty is one which has a great tendency to spontaneous crystallization in any and all parts of the solution. The result in the latter case will be a host of small crystals; in the former, a single large one.

---

[1] Cf. Ostwald, Lehrbuch I, 1036-1039.
[2] Gernez, Beiblätter, **2**, 241 (1878).
[3] Lehrbuch I, 1039.

The method of representing concentrations and temperatures adopted in Fig. 7 is the usual one; but it is not the best. It is possible to represent in the diagram the fusion curve of ice in presence of salt, but not the fusion curve of salt in presence of water, because, at the melting point of the pure salt, the ratio of salt to water becomes infinite and can not be shown in a finite diagram. This can best be done in the manner pointed out by Gibbs,[1] the total mass of the two components, expressed in any units whatsoever, being kept constant. The concentrations can then be represented by points on a horizontal line of definite length and the temperature on an ordinate perpendicular to it. While it is more convenient to plot the concentrations as so many grams of either component in one hundred grams of the mixture because the measurements are made that way and because there are no assumptions made in so doing, it is found better in practice to adopt another scale. It is found experimentally that one gram of one substance is not equivalent chemically to one gram of another substance, and it is more rational to use the chemical unit, the reacting weight, rather than the weight unit, the gram. The concentrations are then expressed in reacting weights[2] of either component per hundred reacting weights of the mixture. In this particular case one hundred and forty-two grams of sodium sulfate are taken as equivalent to eighteen grams of water. The system is in each case supposed to be under its own vapor pressure, at constant volume. Van Rijn van Alkemade[3] prefers to treat the system as if always under constant atmospheric pressure; but this does not seem advisable, as in that case we can not consider the intersection of two curves as representing a nonvariant system. Since the determinations are usually made in open vessels, the system does, as a matter of fact, exist under a constant atmospheric pressure, but that involves bringing in the air as another component. Strictly speaking we do not have equilibrium when working in an open vessel unless the partial pressure of each component in the atmosphere is equal to the partial pressure of the same component in

---

[1] Trans. Conn. Acad. 3, 178 (1876); Cf. Roozeboom, Zeit. phys. Chem. 12, 359 (1893); Konowalow, Wied. Ann. 14, 34 (1881); Alexejew, Ibid. 28, 305 (1886).

[2] Bancroft, Phys. Rev. 3, 25 (1895).

[3] Zeit. phys. Chem. 11, 291 (1893).

equilibrium with the solution and then we are really working at constant volume with an enormously large vapor space. As the change of solubility with the pressure is very slight there will be no experimental difference between the composition of a liquid phase under its own vapor pressure and under a pressure of seventy-six centimeters of mercury.

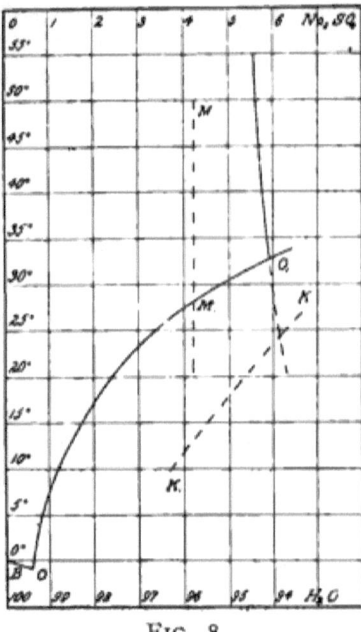

FIG. 8.

In Fig. 8 is shown the equilibrium between water and sodium sulfate. The extreme left of the diagram represents one hundred reacting weights of water and zero reacting weights of sodium sulfate. The letters have the same significance as in Fig. 7, the curves BO, OO$_1$, O$_1$A,[1] KK$_1$ representing the solutions in equilibrium with ice, Na$_2$SO$_4$10H$_2$O, Na$_2$SO$_4$ and Na$_2$SO$_4$7H$_2$O respectively. The curve O$_1$A if prolonged must slant to the right until it cuts the right hand ordinate at the fusion temperature of sodium sulfate, about 860°; but this portion of the curve has not been studied. In this diagram the field in which the unsaturated solution and vapor exist in stable equilibrium lies above the curves BOO$_1$A.

[1] The letter A is missing from Fig. 8.

In the case of sodium sulfate there are only two hydrates known, and one of these represents always a state of labile equilibrium. Neither of these conditions is general and in the equilibrium between calcium chloride and water[1] we have the existence of many hydrates, most of which can exist in stable equilibrium with solution and vapor at some pressures and temperatures. Fig. 9 shows the

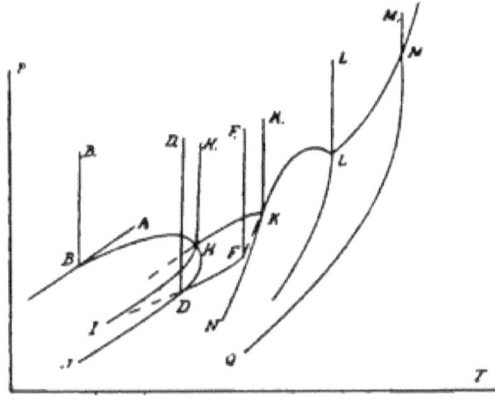

FIG. 9.

pressure-temperature diagram for calcium chloride and water;[2] and Figs. 10 and 11 parts of the same diagram on a larger scale. The ordinates in Fig. 10 are millimeters of mercury; in Fig. 11, centimeters of mercury. The possible solid phases are ice, $CaCl_2 6H_2O$, two modifications of the tetrahydrate $CaCl_2 4H_2O\alpha$ and $CaCl_2 4H_2O\beta$, $CaCl_2 2H_2O$, $CaCl_2 H_2O$ and $CaCl_2$. The equilibrium between anhydrous calcium chloride, solution and vapor, has not been studied in detail owing to the high temperature at which this system first becomes possible. The curves BOHCD, HK, DF, FKL and LM represent the pressures and temperatures at which there can coexist vapor and solutions saturated in regard to $CaCl_2 6H_2O$, $CaCl_2 4H_2O\alpha$, $CaCl_2 4H_2O\beta$, $CaCl_2 2H_2O$ and $CaCl_2 H_2O$ respectively, while the curve AB for the vapor pressures of ice may be assumed to represent the system consisting of ice, solution and vapor. The curves IH, JD, NK and PL represent the systems $CaCl_2 6H_2O$, $CaCl_2 4H_2O\alpha$ and vapor; $CaCl_2 6H_2O$, $CaCl_2 4H_2O\beta$ and vapor; $CaCl_2 4H_2O\alpha$, $CaCl_2 2H_2O$ and vapor; and $CaCl_2 2H_2O$, $CaCl_2 H_2O$ and vapor. The solubility

---

[1] Roozeboom, Rec. trav. chim. **8**, 1 (1889); Zeit. phys. Chem. **4**, 31 (1889).
[2] The letter P is missing from Fig. 9.

of the hexahydrate increases so rapidly with the temperature that the vapor pressure of the saturated solution passes through a maxium at 28.5°, represented by the point O and decreases from there to the point C, at 30.2°, the fusion point of the hexahydrate. At higher

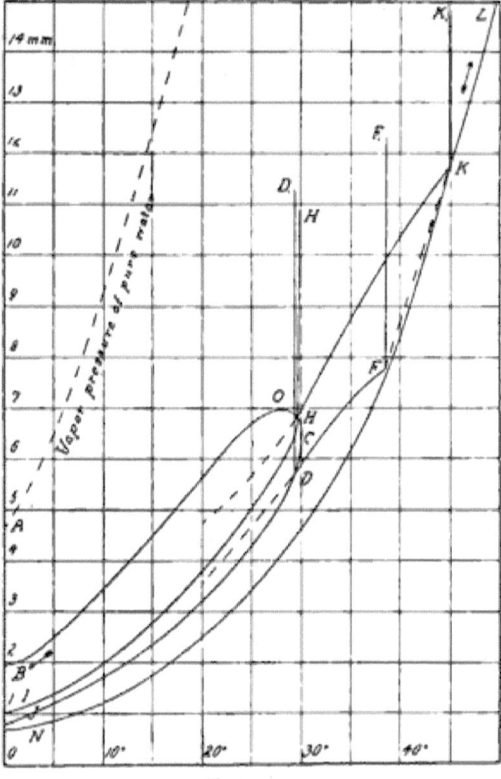

FIG. 10.

temperatures than this the hydrate can not exist; but it can be in equilibrium at lower temperatures with a second solution containing a larger percentage of calcium chloride than the solid crystals. The vapor pressures of these solutions have been determined as far as the point D, this temperature being 29.2°. Strictly speaking, the part of the curve HCD represents a labile equilibrium, the system being instable with respect to $CaCl_24H_2O\alpha$; the solution curve for which, HK, intersects at H, 29.8°, but this modification never appears spontaneously at this point, and there is therefore no difficulty in following the curve BOD to the point D where the other modification with four of water, $CaCl_2 4H_2O\beta$ appears. The saturated solution of $CaCl_2 4H_2O\beta$ is instable with respect to the α modification along its whole length from F (38.4°) to as low temperatures as it has been followed, about 20°. Below 29.2° the solution is also instable

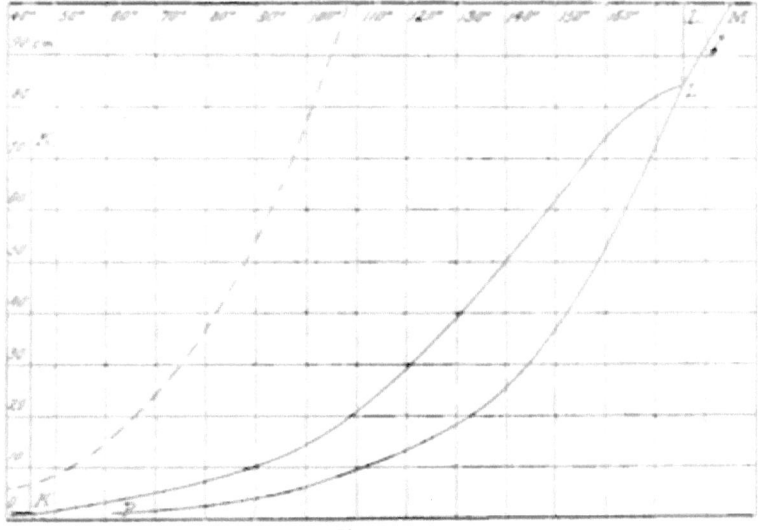

FIG. 11.

with respect to $CaCl_2 6H_2O$. The solution saturated with respect to $CaCl_2 4H_2O\alpha$ can be obtained by the spontaneous change of the β modification and is stable from H to K, from 29.8° to 45.3°. Below the former temperature it is instable with respect to crystals of the hexahydrate; while above the latter the crystals lose water changing into $CaCl_2 2H_2O$. The new curve KL has a maximum vapor pressure at 173°. At 175.5° this salt gives way to $CaCl_2 H_2O$, while the undetermined curve for the anhydrous salt begins at about 260°. When we come to the monovariant systems composed of two solid phases and vapor we find that $CaCl_2 6H_2O$ can exist in equilibrium with either $CaCl_2 4H_2O\alpha$ or $CaCl_2 4H_2O\beta$, there being different pressures in the two cases. This is interesting because it shows the very definite effect exerted by the other solid phase.[1] It seems a truism to say that the equilibrium pressure is different when the sub-

---

[1] Bancroft, Phys. Rev. **3**, 406 (1896).

stances entering into equilibrium change ; but it has not always been clear that the effloresced salt really had any influence. The vapor pressures for the system, $CaCl_2 6H_2O$, $CaCl_2 4H_2O \gamma$ and vapor, are higher than the values at the same temperatures for the system, $CaCl_2 6H_2O$, $CaCl_2 4H_2O \beta$ and vapor, and the latter system is therefore instable with respect to the former. The curve NK represents the stable monovariant system, $CaCl_2 4H_2O \gamma$, $CaCl_2 2H_2O$ and vapor, while the corresponding curve for the system containing $CaCl_2 4H_2O \beta$ instead of the α-modification is too instable to be determined. All that can be said about it is that it will lie below the curve NK except in the immediate neighborhood of the point F. The curve PL for the dihydrate, monohydrate and vapor presents nothing new. The curves for the monovariant systems composed of a liquid and two solid phases were not studied; but they start from the quadruple points, B, H, D, F, K, L and M, and are represented approximately in Fig. 9 by the lines BB_1, HH_1, DD_1, KK_1, LL_1 and MM_1. We shall get a clearer idea of the subject if we consider the concentration-temperature diagram, Fig. 12. The distance between the two chief ordinates represents one hundred formula weights of the mixture, one hundred and eleven grams of calcium chloride being equivalent to eighteen grams of water. The letters have the same significance as in the preceding three figures. Starting at 0° with pure ice and anhydrous calcium chloride, the curve AB represents the temperatures and concentrations at which the monovariant system, ice, solution and vapor exists. At $-55°$ the solution becomes saturated with respect to $CaCl_2 6H_2O$, and we have the nonvariant system, ice, $CaCl_2 6H_2O$, solution and vapor. Further addition of anhydrous calcium chloride causes the ice to disappear with increased formation of the hexahydrate, and the curve BH represents the stable equilibrium between $CaCl_2 6H_2O$, solution and vapor. At 29.8° the solution is saturated with respect to $CaCl_2 4H_2O \alpha$, and if a crystal of this modification be added we shall have the nonvariant system, $CaCl_2 6H_2O$, $CaCl_2 4H_2O \alpha$, solution and vapor. This change does not take place spontaneously, as has been already stated, and it is possible to follow the curve BH through C to D. At this latter temperature, 28.2°, we have the labile but easily realized nonvariant system $CaCl_2 6H_2O$, $CaCl_2 4H_2O \beta$, solution and vapor. At C, 30.2°, we have the fusion point of the hexahydrate. At this temperature the solution has the same composition as the solid phase with which it is in equilibrium. This is, therefore, a true melting point for the hydrate, while the point H corresponds to the inversion point for $Na_2SO_4.10H_2O$. From C to D we have the first example of a solution containing more of a non-volatile solute than the compound which crystallizes from it.

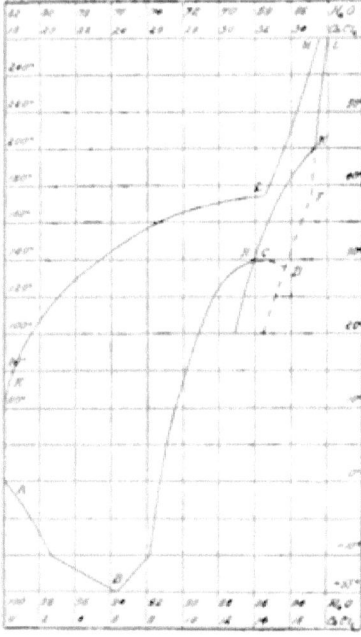

Fig. 12.

This phenomenon had already been observed by Roozeboom for compounds where both the components were volatile.[1] Although the curve CD, like HC, is instable with respect to $CaCl_2.4H_2O$, it is entirely stable with respect to $CaCl_2.6H_2O$, and has been called a stable supersaturated solution. If $CaCl_2.6H_2O$ were to precipitate from it the solution would become more concentrated and would have a lesser vapor pressure. Since the vapor pressure of the solution is the lowest under which $CaCl_2.6H_2O$ can exist at that temperature without efflorescing, it is clear that the hexahydrate can not precipitate without decomposition or, in other words, can not precipitate. The curve DC is to be considered as the prolongation of the curve JD for

[1] Rec. trav. chem., 4, 342 (1885).

the monovariant system, $CaCl_2 6H_2O$, $CaCl_2 4H_2C\beta$ and vapor. The point has been raised whether the curve BHCD is continuous or whether it is composed of two curves, BHC and CD. The former view is supported by Roozeboom,[1] the latter by Le Chatelier.[2] It is still open to any one to take either side. DF is the curve for solutions saturated with respect to $CaCl_2 4H_2O\beta$. Below D, 29.2°, the saturated solution contains more calcium chloride than the solutions saturated at the same temperatures with respect to either $CaCl_2 6H_2O$ or $CaCl_2 4H_2O\alpha$, and is therefore instable with respect to both these salts. From D to F the solution is instable with respect to the less soluble salt, $CaCl_2 4H_2O\alpha$. At F, 38.5°, there can exist the labile nonvariant system, $CaCl_2 4H_2O\beta$, $CaCl_2 2H_2O$, solution and vapor, instable with respect to $CaCl_2 4H_2O\alpha$. The β modification corresponds thus to $Na_2SO_4 7H_2O$ in that it is never in a state of stable equilibrium; but differs qualitatively because the intersections of its solubility curve with the next higher and lower hydrate can be realized experimently. The solubility curve for $CaCl_2 4H_2O\alpha$ has been followed as far as K, 45.3°, where it meets the solubility curve for the dihydrate; in the other direction it has been determined as far as 20°, but the part of the curve beyond H is instable with respect to $CaCl_2 6H_2O$. The solubility curve of the dihydrate is instable with respect to $CaCl_2 4H_2O\alpha$ from F to K. KL represents the stable portion terminated at 175.5° by the appearance of the monohydrate. This latter salt is replaced by the anhydrous salt at the point M, somewhere about 260°, and beyond this the experiments do not go. In Tables XII–XIV are the data for concentrations and pressures at different temperatures. The wavy line in the expression $H_2O \approx xCaCl_2$ denotes a solution containing x reacting weights of calcium chloride dissolved in one of water.[3] The pressures are given in millimeters of mercury and the temperature in Centigrade degrees; x_1 denotes reacting weights of calcium chloride per reacting weight of water; x_2 reacting weights of water per reacting weight of calcium chloride, and x_3, reacting weights of calcium chloride in one hundred reacting weights of the solution.

[1] Comptes rendus, **108**, 744 1013 (1889).
[2] Ibid., **108**, 565, 801, 1015 (1889).
[3] This is not the way Roozeboom writes it, but I have changed his nomenclature to make it agree with his usage in the papers on the gas hydrates.

Table XII

Nonvariant Systems		Temp.	Pressure
Ice,	CaCl$_2$,6H$_2$O,	H$_2$O≕0.069CaCl$_2$, Vapor	− 55°	
CaCl$_2$,6H$_2$O,	CaCl$_2$,4H$_2$Oβ,	H$_2$O≕0.185CaCl$_2$, Vapor	+ 29.2	5.67
CaCl$_2$,6H$_2$O,	CaCl$_2$,4H$_2$Oα,	H$_2$O≕0.164CaCl$_2$, Vapor	29.8	6.80
CaCl$_2$,4H$_2$Oβ,	CaCl$_2$,2H$_2$O,	H$_2$O≕0.207CaCl$_2$, Vapor	38.4	7.88
CaCl$_2$,4H$_2$Oα,	CaCl$_2$,2H$_2$O,	H$_2$O≕0.211CaCl$_2$, Vapor	45.3	11.77
CaCl$_2$,2H$_2$O,	CaCl$_2$,H$_2$O,	H$_2$O≕0.483CaCl$_2$, Vapor	175.5	842.
CaCl$_2$,H$_2$O,	CaCl$_2$,	H$_2$O≕0.556CaCl$_2$, Vapor	260.	

Table XIII

Temp.	Pressure	x_1	x_2	x_3	Temp.	Pressure	x_1	x_2	x_3
Ice, solution and vapor					CaCl$_2$,2H$_2$O, solution and vapor				
0°	4.63	0.000	∞	0.0	40°	8.5	0.208	4.81	17.2
− 5	3.06	0.017	58.7	1.6	45.3	11.8	0.211	4.73	17.4
−10	2.03	0.028	36.3	2.7	50	15.5	0.215	4.66	17.7
−20		0.044	22.8	4.3	55	20.5	0.218	4.59	17.8
−30		0.054	18.4	5.1	60	26.5	0.222	4.51	18.1
−40		0.063	16.0	5.8	65	34.0	0.226	4.43	18.5
−55		0.069	14.5	6.4	70	43.0	0.229	4.37	18.7
CaCl$_2$,6H$_2$O, solution and vapor					75	54.0	0.232	4.31	18.9
−55		0.069	14.5	6.4	80	66.5	0.236	4.24	19.1
−25		0.081	12.3	7.5	85	82.5	0.240	4.16	19.4
−10	0.97	0.089	11.2	8.2	90	100.	0.245	4.08	19.6
0	1.94	0.096	10.4	8.8	100	145.	0.256	3.90	20.4
10	3.46	0.105	9.59	9.6	110	204.	0.269	3.72	21.1
20	5.62	0.121	8.28	10.9	125	326.	0.286	3.50	22.2
25	6.70	0.133	7.52	11.7	135	435.	0.301	3.33	23.1
28.5	7.02	0.147	6.81	12.9	140	497.	0.310	3.23	23.6
29.5	6.91	0.155	6.46	13.5	155	680.	0.348	2.88	25.9
30.2	6.70	0.167	6.00	14.3	160	744.	0.361	2.77	26.5
29.6	5.83	0.175	5.70	14.9	165	790.	0.383	2.61	27.7
29.2	5.67	0.185	5.41	15.6	170	834.	0.413	2.42	29.2
CaCl$_2$,4H$_2$Oα, solution and vapor					175.5	842.	0.483	2.07	32.5
20	4.74	0.148	6.78	12.9	CaCl$_2$,H$_2$O, solution and vapor				
25	5.72	0.156	6.42	13.5	175.5	842.	0.483	2.07	32.5
29.8	6.80	0.164	6.10	14.1	180	910.	0.488	2.05	32.7
35	8.64	0.174	5.75	14.7	185	1006.	0.490	2.04	32.9
40	10.37	0.187	5.34	15.8	190	1114.	0.495	2.02	33.2
45.3	11.77	0.211	4.73	17.4	195	1230.	0.500	2.00	33.3
CaCl$_2$,4H$_2$Oβ, solution and vapor					200	1354.	0.505	1.98	33.5
20	3.56	0.170	5.90	14.5	205	1491.	0.510	1.96	33.8
25	4.64	0.177	5.66	14.9	235		0.538	1.86	35.0
29.2	5.67	0.185	5.41	15.6	260		0.556	1.8	35.7
30	5.83	0.185	5.40	15.7					
35	7.13	0.199	5.04	16.6					
38.4	7.80	0.207	4.83	17.1					

TABLE XIV

Temp.	Pressure CaCl$_2$6H$_2$O,CaCl$_2$4H$_2$Oα Vapor	Temp.	Pressure CaCl$_2$6H$_2$O, CaCl$_2$4H$_2$Oβ Vapor	Temp.	Pressure CaCl$_2$4H$_2$Oα,CaCl$_2$2H$_2$O Vapor	Temp.	Pressure CaCl$_2$2H$_2$O, CaCl$_2$H$_2$O Vapor
−15°	0.27	−15°	0.22	−15°	0.17	65°	842
0	0.92	0	0.76	0	0.59	78	13
10	1.92	10	1.62	10	1.25	100	24
20	3.78	20	3.15	20	2.48	129	60
25	5.08	25	4.32	25	3.40	155	175
29.8	6.80	29.2	5.67	30	4.64	165	438
				35	6.26	170	607
				40	8.53	175.5	715
				45.3	11.77		842

While the system, calcium chloride and water, has brought out several points which did not occur in the system, sodium sulfate and water, such as the stable existence of different hydrates under suitable conditions, the existence of a pressure maximum and of a true fusion point,[1] yet the system CaCl$_2$6H$_2$O, solution and vapor, represents a labile equilibrium at the melting point of the hydrate, and it will therefore be well to study a system in which the melting salt is in stable equilibrium with the solution and vapor. This is the more necessary since the difference between the behavior of the hydrates of sodium sulfate and the hexahydrate of calcium chloride is merely that the labile equilibrium can be realized in the second case and not in the first. The equilibrium between ammonium bromide and ammonia[2] with the three solid phases, NH$_4$Br, NH$_4$BrNH$_3$ and NH$_4$Br 3NH$_3$ illustrates only the points already brought out; but the system, ferric chloride and water,[3] is worth considering in detail. In the concentration-temperature diagram, Fig. 13, three hundred and twenty-five grams of ferric chloride are equivalent to eighteen grams of water.

---

[1] Other salts with true melting points are ZnCl$_2$3H$_2$O, NaH$_2$PO$_4$2½H$_2$O, and the hydrate of Al$_2$Br$_6$. Cf. Roozeboom, Zeit. phys. Chem. **10**, 487 (1892).

[2] Roozeboom, Recueil Trav. Pays-Bas, **4**, 361 (1885); Zeit. phys. Chem. **2**, 460 (1888).

[3] Roozeboom, Zeit. phys. Chem. **10**, 477 (1892).

The solid phases which occur are ice, $Fe_2Cl_6 \cdot 12H_2O$, $Fe_2Cl_6 \cdot 7H_2O$, $Fe_2Cl_6 \cdot 5H_2O$, $Fe_2Cl_6 \cdot 4H_2O$ and anhydrous ferric chloride. AB is the fusion curve with ice as solid phase. At B, $-55°$, the solution becomes saturated with respect to $Fe_2Cl_6 \cdot 12H_2O$, forming the nonvariant system, ice, $Fe_2Cl_6 \cdot 12H_2O$, solution and vapor. BCDN is the solubility curve for $Fe_2Cl_6 \cdot 12H_2O$. At C, $37°$, the solution has the same composition as the hydrate, and this temperature is the melting point of $Fe_2Cl_6 \cdot 12H_2O$. From C to N the saturated solution

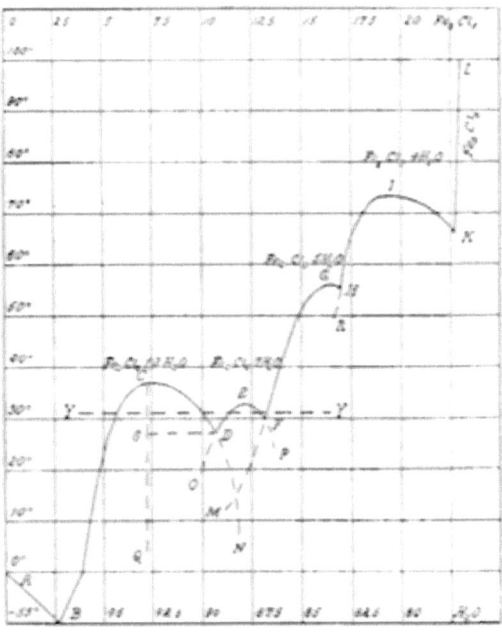

FIG. 13.

contains more ferric chloride than the crystals. As the hydrate, $Fe_2Cl_6 \cdot 7H_2O$, can appear at D, $27.4°$, the part of the curve DN represents a state of labile equilibrium, instable with respect to the hydrate with seven of water. At D there can coexist, $Fe_2Cl_6 \cdot 12H_2O$, $Fe_2Cl_6 \cdot 7H_2O$, solution and vapor. The curve ODEFP is the solubility curve for $Fe_2Cl_6 \cdot 7H_2O$, the parts OD and FP representing labile equilibrium. The hydrate melts at $32.5°$, shown in the diagram by the point E. At F, $30°$, there is the monovariant system, $Fe_2Cl_6 \cdot 7H_2O$, $Fe_2Cl_6 \cdot 5H_2O$, solution and vapor. MFGH is the solubility curve for the hydrate $Fe_2Cl_6 5H_2O$, with a fusion point at G, 56°. The labile portion of this curve, FM, intersects the curve DN at about 15°, and at this temperature there can exist the labile nonvariant system, $Fe_2Cl_6 12H_2O$, $Fe_2Cl_6 5H_2O$, solution and vapor, instable with respect to $Fe_2Cl_6 7H_2O$. At 55°, H, there exists the stable nonvariant system with $Fe_2Cl_6 5H_2O$ and $Fe_2Cl_6 4H_2O$ as solid phases and at K, 66°, another with $Fe_2Cl_6 4H_2O$ and anhydrous ferric chloride. RHIK represents the solubility curve for $Fe_2Cl_6 4H_2O$, the fusion point I coming at 73.5°. From K to L the solution is in equilibrium with anhydrous salt; but this curve has not been followed beyond 100°, owing to the partial decomposition of the ferric chloride. In the case of ferric chloride and water, there are four hydrates, each of which can exist in stable equilibrium with solutions containing more salt than the crystals and, in consequence, each of the four is in stable equilibrium at its melting point. It is worth noticing that the hydrate, $Fe_2Cl_6 7H_2O$, is in stable equilibrium only over a very narrow range of temperatures and concentrations, and possibly never would have been discovered except by a careful study of the solubility curves. As a matter of fact, Roozeboom noticed that solutions saturated with respect to $Fe_2Cl_6 5H_2O$ sometimes solidified entirely at the temperature and concentration corresponding to F, while solutions saturated in respect to $Fe_2Cl_6 12H_2O$ showed the same phenomenon under the circumstances represented by D. From the Phase Rule he knew it was impossible that the solid phases should be the same in the two cases, and he was thus led to discover the hydrate with seven of water. In this particular case the two temperatures are nearly three degrees apart, but it is conceivable that this difference might be zero, and then only the solubility determinations would show the existence of a hitherto unknown compound. The diagram, Fig. 13, enables us to predict the behavior of a ferric chloride solution if evaporated to dryness at a constant temperature of about 31°. The solution would first solidify to $Fe_2Cl_6 12H_2O$, become liquid, solidify to $Fe_2Cl_6 7H_2O$, become liquid yet again and solidify for the third time with formation of $Fe_2Cl_6 5H_2O$. Further abstraction of water would cause this last hydrate to effloresce in the usual manner. This will be clear if one follows the horizontal dotted line YY. All these different solids and solutions represent states of stable equilibrium. In other words, the portion of the line YY which is bounded by EF and FG represents a solution stable with respect to $Fe_2Cl_6 7H_2O$ and to $Fe_2Cl_6 12H_2O$, although it contains more of the solute than the solutions saturated with respect to either of these salts. It is evidently not possible to define a supersaturated solution as one which contains more of the solute than the saturated solution, although this definition would have applied to all the solutions studied up to now, with the exception of those mixtures of calcium chloride and water in the field bounded by CHD and an isothermal line through C in Fig. 12. These solutions are stable with respect to $CaCl_2 6H_2O$. As almost the whole of this field is instable with respect to $CaCl_2 4H_2O$ the point was not mentioned in the discussion of the equilibrium between calcium chloride and water, it being reserved until a more striking illustration of the phenomenon could be found.

All solutions between the temperatures represented by C and D, Fig. 13, which contain more ferric chloride than the point on CD for that temperature have lower pressures than the minimum pressure at which $Fe_2Cl_6 12H_2O$ can exist, and it is therefore impossible for this salt to crystallize from the solution. The same reasoning applies to the part of the line YY bounded by EF and FG, except that here the vapor pressures are lower than the minimum pressure for either $Fe_2Cl_6 12H_2O$ or $Fe_2Cl_6 7H_2O$, and consequently neither can crystallize from the solution. These are cases of what have been called, for lack of a better name, stable supersaturated solutions.[1] It does not follow that a hydrate can never separate from a solution which has a lower vapor pressure than the minimum for that hydrate. This is true only in case the solution would become more concentrated by the crystallization of the hydrate; in other words, when the solution contains more of the solute than the solid phase in question. If the word "supersaturated" be used in the normal sense to signify a solution instable in presence of the solid phase, with respect to which it is said to be supersaturated, the only solutions supersaturated in respect to $Fe_2Cl_6 7H_2O$ are those in the field OEP, while those saturated in respect to $Fe_2Cl_6 12H_2O$ are bounded by the curve BCN.[2] If

---
[1] Bancroft, Jour. Phys. Chem. **1**, No. 3 (1896).
[2] Roozeboom, Zeit. phys. Chem. **10**, 492 (1892).

CQ be a line perpendicular to the abscissae, all solutions to the left of it contain less ferric chloride and all to the right of it more ferric chloride than the hydrate $Fe_2Cl_6 12H_2O$. Let DS be an isothermal line passing through D and meeting CQ at S. It is then possible to predict the behavior of any solution supersaturated with respect to $Fe_2Cl_6 12H_2O$, when a crystal of this hydrate is added, the system being kept always at constant temperature. From any solution in QCB the system passes to the corresponding solution on the curve BC; from any point in the field DSC, it passes to the corresponding point on DC. In both these cases the monovariant system formed, $Fe_2Cl_6 12H_2O$, solution and vapor, is entirely stable. From any point in the field NDSQ there will be, as before, precipitation of $Fe_2Cl_6 12H_2O$ and formation of the system hydrate, solution and vapor; but the solution is in a state of labile equilibrium, because the curve DN lies within the field OEP for solutions supersaturated with respect to $Fe_2Cl_6 7H_2O$. The final stable equilibrium for a solution having the composition represented by a point in the field NDSA will be the system, $Fe_2Cl_6 12H_2O$, $Fe_2Cl_6 7H_2O$, and vapor. If the original solution were represented by a point in the field NDO, a crystal of either $Fe_2Cl_6 12H_2O$ or $Fe_2Cl_6 7H_2O$ would cause a precipitation of the salt added, because the solution is supersaturated with respect to both hydrates; in the other parts of the field NDSQ the hydrate with twelve of water must crystallize first, and addition of solid $Fe_2Cl_6 7H_2O$ will have no effect. Similar phenomena take place in the supersaturated regions of the other hydrates. From its position as the last hydrate before the appearance of solid solvent, solutions supersaturated with respect to $Fe_2Cl_6 12H_2O$ can be supersaturated only with respect to $Fe_2Cl_6 7H_2O$, while the supersaturated region OEP for $Fe_2Cl_6 7H_2O$ overlaps that for $Fe_2Cl_6 12H_2O$, in the part NDO, and that for $Fe_2Cl_6 5H_2O$, in the part PFM. Bearing this in mind, there is no difficulty in applying all that has been said about solutions supersaturated with respect to $Fe_2Cl_6 12H_2O$ to solutions supersaturated with respect to the other hydrates. The best definition of a supersaturated solution is one by Budde, quoted by Roozeboom in his paper on ferric chloride.[1] "A solution is supersaturated with respect to a definite solid phase if, in contact with it at the temperature of

---

[1] Zeit. phys. Chem. **10**, 495 (1892).

the experiment, more of that phase separates. Such a solution contains more salt than the saturated solution if the latter contains more water at that temperature than the solid phase, and *vice-versa*." This is put in the following form by Roozeboom: "A solution is supersaturated with respect to a solid phase at a given temperature if its composition is between that of the solid phase and the saturated solution."

Since the solution at D is more concentrated than $Fe_2Cl_6 12H_2O$ and less so than $Fe_2Cl_6 7H_2O$, it follows that this solution will solidify to the two hydrates without change of temperature. This occurs only in case the composition of the solution lies between those of the two solid phases and is not a general phenomenon. The solution in equilibrium with $Na_2SO_4 10H_2O$ and Na_2SO_4 contains more water than either salt, and the same is true of the solution in equilibrium with $CaCl_2 4H_2O\alpha$ and $CaCl_2 2H_2O$. In both these cases the solution will deposit, on cooling, the salt stable at lower temperatures and the temperature will fall.

In Tables XV-XVI are the experimental data for ferric chloride and water. Under the heading x_1 are the concentrations in reacting weights of ferric chloride per reacting weight of water; under x_2 the reciprocal values, and under x_3 the reacting weight of ferric chloride in one hundred reacting weights of the solution. Under the heading "nonvariant systems," only the solid phases are given, solution and vapor being understood in each case.

Table XV

Nonvariant Systems		Temp.
Ice,	$Fe_2Cl_6 12H_2O$	$-55°$
$Fe_2Cl_6 12H_2O$,	$Fe_2Cl_6 7H_2O$	27.4
$Fe_2Cl_6 7H_2O$,	$Fe_2Cl_6 5H_2O$	30.
$Fe_2Cl_6 12H_2O$,	$Fe_2Cl_6 5H_2O$	15.
$Fe_2Cl_6 5H_2O$,	$Fe_2Cl_6 4H_2O$	55.
$Fe_2Cl_6 4H_2O$,	Fe_2Cl_6	66.
Melting Points		
$Fe_2Cl_6 12H_2O$		37.
$Fe_2Cl_6 7H_2O$		32.5
$Fe_2Cl_6 5H_2O$		56.
$Fe_2Cl_6 4H_2O$		73.5

TABLE XVI.

Temp.	x_1	x_2	x_3	Temp.	x_1	x_2	x_3
Ice, solution and vapor				$Fe_2Cl_6 5H_2O$, solution and vapor			
0°	0.000	∞	0.00	12°	0.129	7.77	11.17
−10	0.010	100.	0.99	20	0.139	7.17	12.25
−20.5	0.016	61.	1.62	27	0.149	6.73	12.94
−27.5	0.019	52.6	1.87	30	0.151	6.61	13.12
−40	0.024	42.2	2.32	35	0.156	6.40	13.52
−55.	0.027	36.4	2.65	50	0.175	5.71	14.90
$Fe_2Cl_6 12H_2O$, solution and vapor				55	0.192	5.22	16.07
−55	0.027	36.4	2.65	56	0.200	5.00	16.67
−41	0.028	35.6	2.74	55	0.203	4.92	16.91
−27	0.030	33.6	2.89	$Fe_2Cl_6 4H_2O$, solution and vapor			
0	0.041	24.2	3.97	50	0.200	5.01	16.54
10	0.045	22.0	4.35	55	0.203	4.92	16.91
20	0.051	19.6	4.85	60	0.207	4.88	17.14
30	0.059	16.9	5.60	60	0.215	4.64	17.70
35	0.068	14.8	6.35	72.5	0.234	4.28	18.50
36.5	0.079	12.6	7.34	73.5	0.250	4.00	20.00
37	0.083	12.0	7.69	72.5	0.262	3.82	20.70
36	0.093	10.8	8.49	70	0.279	3.58	21.83
33	0.105	9.57	9.25	66	0.292	3.43	22.60
30	0.112	8.92	9.84	Fe_2Cl_6, solution and vapor			
27.4	0.122	8.23	10.84	66	0.292	3.43	22.60
20	0.128	7.80	11.38	70	0.294	3.40	22.7
10	0.132	7.57	11.67	75	0.289	3.46	22.45
8	0.137	7.30	12.05	80	0.292	3.43	22.6
$Fe_2Cl_6 7H_2O$, solution and vapor				100	0.298	3.33	22.9
20	0.114	8.81	10.19				
27.4	0.122	8.23	10.84				
32	0.136	7.38	11.94				
32.5	0.143	7.00	12.50				
30	0.151	6.61	13.12				
25	0.155	6.47	13.45				

While it has been possible to realize some states of labile equilibrium in the systems which have been under discussion, it has not been easy to follow the curves more than a short distance beyond the point where the new phase could appear. Systems differ very much in respect to this. It has already been mentioned that with the divariant systems, solution and vapor, it is possible to obtain marked supersaturation with some solutes, almost impossible with others. In the same way, some monovariant systems change spontaneously as soon as the equilibrium for the nonvariant system is reached, while others can be carried far beyond it. A striking example of the latter is to be found in the hydrates of thorium sulfate, the behavior of which has been studied by Bakhuis Roozeboom.[1] Fig. 14 is the con-

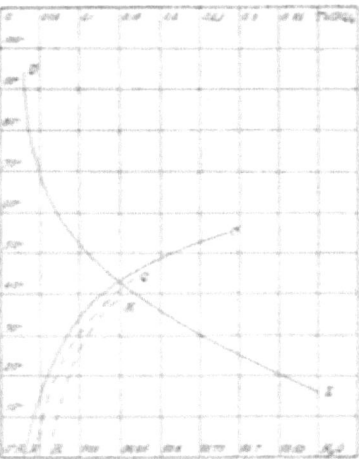

FIG. 14.

centration-temperature diagram for thorium sulfate and water. ABC, FG, HK and DBE[2] represent the solutions saturated with respect to thorium sulfate with nine, seven, six and four of water, respectively. It is clear that AB and BD are the only stable solutions; but the labile forms can be obtained over a wide range of temperatures. Even in presence of the stable modification it is often hours and even days before the change is completed. If the anhydrous salt be brought in contact with water, it will often dissolve completely, only precipitating the stable hydrate after long standing, whereas it is usual for a dehydrated salt to take up water at once.[3]

[1] Zeit. phys. Chem. 5, 198 (1890).
[2] Th(SO$_4$)$_2$.4H$_2$O dissolves with evolution of heat. The letter B is missing from the intersection of AC and DE.
[3] The hydrates of manganous sulfate seem to present another case of extraordinary sluggishness; but the statements with respect to them are so improbable as to render the paper valueless. Linebarger, Am. Chem. Jour. 15, 225 (1893).

CHAPTER VI

VOLATILE SOLUTES

It has been assumed hitherto that one of the components had an immeasurably low vapor pressure. Under these circumstances the vapor phase contains only one component, and condensation has the effect of adding more of that component, of diluting the solution if there happens to be a liquid phase. If both components are volatile this is no longer true, and under certain circumstances condensation means adding more of one component, and under others the reverse. With one non-volatile component, increase of external pressure causes the monovariant system, solid, solution and vapor, to pass into one of the two divariant systems, solid and solution, or solution and vapor, except when the solution contains more of the solute than the solid phase. When both components are volatile this is no longer so, and it is possible to pass by increased pressure from solid, solution and vapor to solid and vapor. This may be illustrated by the equilibrium between iodine and chlorine.[1] The solid phases which occur are iodine, two modifications of iodine monochloride and iodine trichloride. Under suitable circumstances, it would be possible to have solid chlorine; but this does not come within the range of the present discussion. The vapor pressures of the labile monochloride $ICl\beta$ differed so little from those of $ICl\alpha$ that they could not be determined and are, therefore, not given in the pressure-temperature diagram, Fig. 15. This equality applies only to the total pressure. It is not probable that the partial pressures of iodine and chlorine vapor in equilibrium with the two modifications are the same, but only that their sum is. This point has not been determined experimentally.

In the diagram AB represents I_2, $ICl\alpha$ and vapor; BC, I_2, solution and vapor; BDE, $ICl\alpha$, solution and vapor; HE, $ICl\alpha$, ICl_3 and vapor; EF, ICl_3, solution and vapor.[2] The monovariant systems

---

[1] Stortenbeker, Zeit. phys. Chem. **3**, 11 (1889); **10**, 183 (1892).
[2] In order to keep the figure a convenient size, two pressure scales are used.

containing two solid phases and solution have not been studied ; but they may be represented approximately by the dotted lines BB_1 and EE_1. The nonvariant systems are I_2, $ICl\alpha$, solution and vapor existing at B, 7.9°, and $ICl\alpha$, ICl_3, solution and vapor possible at E, 22.7°. The curve C_1C represents the vapor pressure of solid iodine, terminating at 114.2°, the melting point of pure iodine, CC_2 being the curve for liquid iodine and vapor. Below this curve solid iodine can not exist. It will be noticed that the vapor pressure of the system

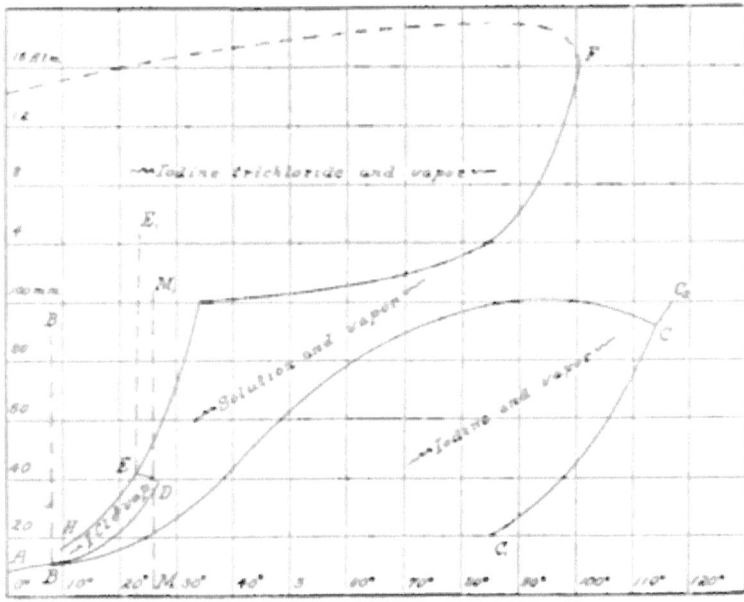

FIG. 15.

iodine, solution and vapor, is higher than that of the solid solvent, a phenomenon which can only occur with a volatile solute. In the neighborhood of 100° this curve has a maximum vapor pressure analogous to the maximum vapor pressure of the system, $CaCl_2 6H_2O$, solution and vapor, just below the fusion temperature of the hexahydrate. At 63.7° the vapor pressure of the system, ICl_3, solution and vapor, becomes equal to atmospheric pressure so that the melting point of the compound can not be determined in open vessels. There exist in the fields C_1CBA, ABDEH, HEF and FEDBC, the divariant systems composed of vapor in equilibrium with iodine, monochloride, trichloride and solution, respectively. As these are the more interesting systems I have marked them on the diagram. The other divariant systems, iodine and monochloride, iodine and solution, monochloride and trichloride, monochloride and solution, trichloride and solution, exist in the fields B_1BA, B_1BC, E_1EH, B_1BDEE_1, E_1EF.

If we start with a very large quantity of iodine and chlorine vapors at the pressure and temperature represented by the point M, and compress the mixture at constant temperature, the changes will be represented by the dotted line MM_1. The following monovariant and divariant systems will be formed: iodine and vapor; iodine, solution and vapor; solution and vapor; monochloride, solution and

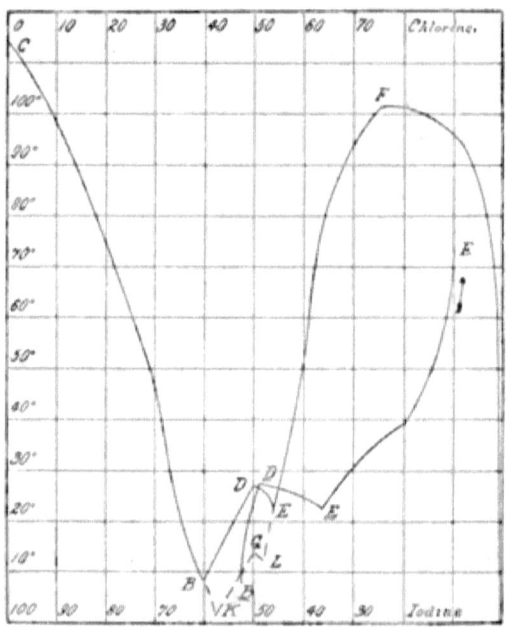

FIG. 16.

vapor; monochloride and vapor; monochloride, solution and vapor; solution and vapor; trichloride, solution and vapor; trichloride and vapor; trichloride, solution and vapor; solution and vapor. This multitude of changes is due largely to the fact that both the monochloride and the trichloride can exist in stable equilibrium with solutions containing more and less chlorine than the crystals themselves. This, as well as the place occupied by ICIβ in the equilibrium, will be clearer if we consider the compositions of the different solutions as shown in the concentration-temperature diagram, Fig. 16. One hundred and twenty-seven grams of iodine are equivalent to thirty-five and one-half grams of chlorine. The composition of the vapors is represented as far as possible. The letters have the same significance as in the preceding diagram. The curves for the vapors have the same letters as the corresponding solutions, but underlined to distinguish them. The melting point of ICIα is found to be at $27.2°$, KGL is the solubility curve for ICIβ, this compound melting at $13.9°$. The curve represents a state of labile equilibrium at all temperatures. There is a labile nonvariant system possible at K, $0.9°$, and another at L, $12.0°$. At the first there coexist I_2, ICIβ, solution and vapor; at the second, ICIβ, ICI$_3$, solution and vapor. The continuation of the curve EF represents the solutions in equilibrium with ICI$_3$, and containing more chlorine than the crystals. This curve if prolonged terminates at about $-102°$, when solid chlorine begins to separate from the solution. There seems to be no question from the experimental data that the two branches of the solubility curve for ICIα, BD and ED meet at an angle, but this is denied by Stortenbeker on the strength of a thermodynamic formula by van der Waals. In view of the tacit assumptions which have been so often discovered in thermodynamical formulas, this conclusion can hardly be considered a wise one, especially as we are not yet in a position to say at what point iodine ceases to be solvent and becomes solute, nor what effect this will have on the equations of equilibrium. In this particular case it is assumed explicitly[1] that the solution consists of a mixture of iodine and chlorine, and that there is no fused monochloride present. This may or may not be true. There is no way at present of determining it. In Tables XVII–XVIII are the experimental data for this system. The pressures are given in millimeters of mercury. x_1 denotes units of chlorine per unit of iodine; x_2, units of iodine per unit of chlorine; x_3, units of chlorine in one hundred units of solution; y, units of chlorine per unit of iodine in the vapor. The values of y are only approximative.

[1] Zeit. phys. Chem. **10**, 194 (1892).

TABLE XVII

Nonvariant Systems		Vapor	Temp.	Pressure
Solid	Liquid			
I_2, IClα	I ≈ 0.66 Cl	I ≈ 0.92 Cl	7.9°	11
I$_2$Clα, ICl$_3$	I ≈ 1.19 Cl	I ≈ 1.75 Cl	22.7	42

Melting Points

I_2, 114.2°	IClα, 27.2°	IClβ, 13.9°	ICl$_3$, 101°

TABLE XVIII

Temp.	Pressure	x_1	x_2	x_3	y	Temp.	Pressure	x_1	x_2	x_3	y
Iodine, solution and vapor.						IClα, ICl$_3$, vapor.					
5.0°		0.69	1.45	41		9.7°	16.				
7.9	11.	0.66	1.51	40		15.	24.				
10.		0.65	1.54	39		19.	32.				
20.	15.	0.54	1.85	35		20.	36.				
25.	20.					22.7	42.				
30.	25.	0.49	2.04	33		ICl$_3$, solution and vapor.					
40.	43.	0.45	2.22	31		20.		1.17	0.85	54.	
50.	63.	0.40	2.50	29		22.7	42.	1.19	0.84	54.	1.75
70.		0.28	3.57	22		25.	49.5				
100.		0.10	10.	9		30.	72.	1.26	0.79	56.	2.36
114.2	91.	0.00	∞	0		40.	147.	1.37	0.73	58.	4.20
IClα, solution and vapor.						50.	296.				
7.9	11.	0.66	1.51	40	0.92	60.	571.	1.55	0.65	61.	7.0
10.	12.	0.69	1.45	41		64.1	773.				
13.5	13.5					70.1	1183.				
15.	16.	0.76	1.31	43	0.93	73.6		1.66	0.60	63.	8.7
20.	22.	0.84	1.19	46		78.7	2284.				
22.5	24.5					85.3	3549.				
27.2	39.	1.00	1.00	50	1.04	89.		2.02	0.50	67.	
26.0	41.					90.4	5190.				
25.	41.5	1.11	0.90	53	1.45	95.4	8137.				
22.7	42.	1.19	0.84	54	1.75	96.		2.43	0.41	71.	
IClβ, solution and vapor.						100.5	11707.				
0.9		0.72	1.42	42		101.	16 Atm.	3.00	0.33	75.	
7.0		0.84	1.19	46		94.				92.	
13.9		1.00	1.00	50		75.				97.7	
12.0		1.10	0.91	52		60.5				99.	
Iodine, IClα and vapor.						42.5				99.3	
5.0	9.					30.				99.8	
7.9	11.										

It often happens that a volatile solute forms no solid compounds with the solvent and, in that case, the only monovariant system possible will be solid solvent, solution and vapor, provided the pressure be not increased to the point where the solute condenses to a liquid, and provided the solvent occurs in only one solid modification. The diagram for such a system, the solute being a gas, is confined to the fusion curve CB, Fig. 16. For each pressure there is a definite concentration in the solution phase and a definite temperature at which the monovariant system can exist. The pressure, being the sum of the partial pressures of the solvent and solute, is usually higher than that of the solid solvent at the same temperature; but with a very soluble gas this is not necessarily true, and the system may behave like one with a non-volatile solute, or the vapor pressure of the system may even pass through a minimum.

An example of this is possibly to be found in the equilibrium between hydrochloric or hydrobromic acid and water, though this particular curve has not been studied by Roozeboom.[1] Neither of these instances is really satisfactory, since both hydrochloric and hydrobromic acid form solid compounds with water. With a sparingly soluble gas the increase of pressure with increasing concentration may be very great, and it becomes an interesting question whether this may not have some effect on the equilibrium between solid and solution, considered as an increase of pressure. The subject has not been studied carefully from this point of view, but the bulk of the evidence points to the conclusion that the equilibrium between solid and solution or solid and solid[2] is a function of the total pressure on the solid phase, independent of the fact that this pressure is due in part to one of the components in the vapor phase.[3] There will be two influences at work to change the freezing point in the case of a system composed of solid solvent, solution and vapor. There will be the change in the partial pressure of the solvent due to the concentration of the solute in the solution phase. This always lowers the freezing point when the solid phase is pure solvent. In addition

---

[1] Recueil Trav. Pays-Bas, **3**, 84 (1884); Zeit. phys. Chem. **2**, 454 (1888).
[2] Cf. Reicher, Recueil Trav. Pays-Bas, **2**, 246 (1883).
[3] For numerical data in regard to the freezing point **of water when saturated with** different gases at atmospheric pressure, see Prytz, Beiblätter **19**, 870 (1895).

there is the increased pressure due to the concentration of the solute in the vapor phase. This will lower the freezing point if the solvent is less dense as solid than as liquid, which is the case with water; it will raise the freezing point in case the solvent contracts in freezing, as most solvents do. In the first case the two forces work together giving a greater depression of the freezing point than would be calculated from the concentration alone; in the second case they act in opposite directions. It is quite possible that with a substance which expands in melting and a gas almost insoluble in the liquid solvent, the effect due to pressure might eventually overbalance the effect due to solubility; in which case the freezing point would sink, pass through a minimum and then increase. This has never been observed experimentally. The hypothetical case that the volume change might have different signs when the solvent separates as solid from the pure liquid or from the solution, is too complicated to be worth considering.

Of more importance is the behavior of the divariant system, solid and vapor. Here there can be any pressure at any temperature within the limits for the appearance of new phases; but the question suggests itself whether the second or gaseous component has any effect upon the partial pressure of the solid. It is held that there is no such influence;[1] but this does not seem entirely satisfactory.[2] There are cases of increased volatility on the part of the solid which do not seem to be explicable by assuming the formation of volatile compounds. Instances of this are the behavior of iodine in the presence of carbonic acid;[3] of potassium iodide in alcohol vapor;[4] of boric acid and methyl alcohol;[5] of zinc oxide and sulfite in presence of zinc vapor.[6] Instances where there may easily be formation of volatile compounds but where it is doubtful whether that explana-

[1] Ostwald, Anal. Chem. 33; Nernst, Theor. Chem. 306.
[2] Bancroft, Jour. Phys. Chem. 1, No. 3 (1896).
[3] Villard, Comptes rendus, 120, 182 (1895).
[4] Hannay, Proc. Roy. Soc. 30, 178 (1880).
[5] Gooch, Proc. Am. Acad. 22, 167 (1886). Cf. also Margueritte-Delacharlonnay, Comptes rendus, 103, 1128 (1886); Bailey, Jour. Chem. Soc. 65, 445 (1894).
[6] Morse and White, Am. Chem. Jour. 11, 258, 348 (1889).

tion is sufficient are the volatility of nickel chloride in a current of hydrogen and of iron, nickel and zinc in a current of hydrochloric acid.[1] In these last cases we have more than two components. The volatility of platinum in presence of chlorine, hydrogen or nitrogen[2] under proper circumstances is a very interesting phenomenon though very possibly having no bearing on the question under consideration. The point at issue is whether a vapor phase containing two or more components is a solution or a mixture. If the change from a liquid to a vapor is a continuous one, the vapor must possess a solvent power though it may be to an infinitesimal degree. If we accept this we are forced to conclude that a vapor or gas always has an effect upon the partial pressure of a solid with which it forms no compounds. The partial pressure of the solid will increase if the solid is soluble in the other vapor or gas and will decrease if the reverse is true. While this effect may fall within the limits of experimental error in many cases, it is extremely improbable that it does so always.

A different nonvariant system from any yet considered is one in which there is equilibrium between two solid modifications of the solute, solution and vapor. Sulfur crystallizes in the monoclinic form from a hot solution in toluene, in the rhombic form at lower temperatures. From this it follows that at some intermediate temperature the two modifications can exist simultaneously in stable equilibrium with the solution and vapor. So far as we know this temperature is identical with the inversion point when no solvent is present.[3]

---

[1] Schützenberger, Comptes rendus, **113**, 177 (1891).
[2] Lehmann, Molekularphysik, II, 72.
[3] Cf. Nernst, Theor. Chem. 505.

CHAPTER VII

TWO LIQUID PHASES

There are many substances which are not miscible in all proportions in the liquid state, and, with two components of this class, it is possible to have two liquid phases, one containing an excess of one of the substances, the other of the other. The component which is in excess is in all cases the solvent, the other the solute and it is customary to speak of one of the liquid phases as a solution of A in B and of the other as a solution of B in A. As an example of this class we will take naphthalene and water. Fig. 17 gives an approximate representation of the change of concentration with the temperature. The diagram lays no claim to accuracy because the solubilities have not been determined. AB is the fusion curve for naphtha-

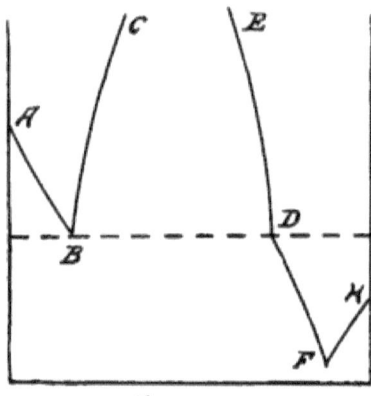

FIG. 17.

lene in the presence of water. Naphthalene is the solvent and water is the solute, the latter lowering the freezing point of the former. When the concentration represented by B is reached, further addition of water brings about no further depression of the freezing point. Instead, there appears a second liquid layer having the concentration represented by the point D, forming the nonvariant system, solid naphthalene, solution of water in naphthalene, solution of naphthalene in water, and vapor. The temperature at which this is possible is about 74°. This system can exist only at one temperature and one pressure. Any change in the conditions of equilibrium brings about the disappearance of one of the phases before either temperature or pressure can change. If the solid phase be made to disappear we have the monovariant system, two liquid phases and vapor. For a given temperature the pressure of the system and the concentrations in the two liquid phases are entirely determined and addition of either water or naphthalene produces a change in the relative masses of the phases, not in their compositions. These latter are represented by the curves BC and DE. It would be possible to take either curve as representing the monovariant system and disregard the other but it conveys more information to have both in the diagram, and the points at which an isothermal line cuts the two curves show the concentrations of the two solutions in equilibrium.

Since the two solutions and the vapor constitute a monovariant system they must have a constant boiling point so long as the external pressure remains constant. This is found to be true experimentally, the boiling mixture distilling at constant temperature—in this case at about 95°—so long as the two liquid phases are present.

It is to be noticed that there are two sets of solutions represented by BA and BC in both of which naphthalene is solvent. The first is a fusion curve, the second a solubility curve. The distinction between them is that in the latter case the solution is saturated in respect to the solute, in the former in respect to the solvent.

If a mixture of naphthalene and water forming two liquid phases and vapor be allowed to cool at constant volume, the temperature will fall until the solid naphthalene begins to appear at 74°. The temperature will remain constant until the whole of the liquid phase, solution of water in naphthalene, has disappeared leaving the monovariant system, solid naphthalene, solution of water in naphthalene, and vapor. The temperature falls again, the system passing along the line DF until at F the cryohydric point is reached and ice separates, forming the nonvariant system, naphthalene, ice, solution of naphthalene in water and vapor. The temperature now remains constant until the whole of the solution has disappeared. DF is a continuation of ED, the break in the curve at D being caused by the change in the heat of solution when the solute separates in the solid state.[1] In the solutions represented by ED and DF, water is solvent and naphthalene is solute. The curve FH is the fusion curve for ice. The complete concentration-temperature diagram for two substances, which do not form any compounds and which do form two liquid phases at some temperature, consists of two fusion curves and two solubility curves. In the fields ABC and EFH there exist the divariant systems, solution and vapor. In the first, naphthalene is solvent; in the second, water. These systems may have either higher or lower vapor pressures than the pure solvents. The relation of these vapor pressures to those of the monovariant systems, two liquid phases and vapor is made clear by following the change of pressure with the concentration at constant temperature. In the concentration-pressure diagram, Fig. 18, are four isothermal curves representing the behavior of typical pairs of partially miscible liquids.[2]

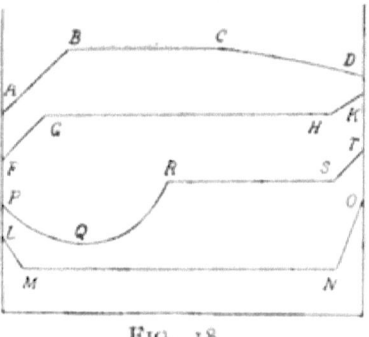

FIG. 18

AB shows the pressures of the divariant system, solution and vapor, with the less volatile component as solvent. At B the solution becomes saturated in respect to the second component. BC is the vapor pressure of the monovariant system with two liquid phases. It is a horizontal line because the pressure remains constant at constant temperature so long as the two liquid phases are present, in-

---

[1] Walker, Zeit phys. Chem. **5**, 192 (1890).
[2] Cf. Konowalow, Wied. Ann. **14**, 219, (1881).

dependent of the percentage composition of the mixture. In passing from B to C the concentrations of the two liquid phases remain constant for the same reason that the pressures do and the only difference is in the relative quantities of these two phases. At B there is practically none of the solution in which the second component is solvent; at C none of the solution in which the first component is solvent. Since one of the liquid phases disappears at C, a further increase in the relative amount of the second component leads to the formation of unsaturated solutions and the pressure curve CD, for these divariant systems ends at D with the vapor pressure of the second component. The characteristic feature of this curve is that the vapor pressure of the monovariant system, represented by BC, is higher than that of either of the single components. This curve is typical of most pairs of partly miscible liquids, such as mixtures of water with chloroform, toluene, benzene, ether or naphthalene and it was thought at one time that probably no other form of curve occurred. The points B and C which represent the pressures and concentrations of the two saturated solutions will move nearer each other as the liquids are more soluble one in the other. For the hypothetical case that the two liquids are absolutely non-miscible the two side curves AB and CD will disappear and the line BC will represent the pressures of all the mixtures. Under these circumstances the vapor pressure of the system will be the sum of the vapor pressures of the pure components. This case is realized very nearly by the system, chloroform and water.[1] As the mutual solubilities of the two liquids increase, the vapor pressure of the monovariant system falls farther and farther below the sum of the pressures of the two pure components. If the vapor pressure of one of the components is low in comparison with that of the other, it may readily happen that the decrease in the vapor pressure of the solvent when the less volatile component is solute may be larger than the partial pressure of the solute. This is the normal case with solutes which are solids in the pure state at the temperature of the experiment and may be expected with all liquid solutes when the ratio of the vapor pressures of the two components falls below a cer-

---

[1] Ostwald, Lehrbuch I, 641.

tain value which is a function of the reacting weights of the pure components.[1] The isothermal vapor pressure curve for such a system is represented by FGHK, Fig. 18. FG and HK are the curves for the two unsaturated solutions while the two liquids and vapor exist along GH. The difference between this case and the more usual one is that the vapor pressure of the monovariant system has a value between those of the two components. The first example of such an equilibrium was observed by Roozeboom[2] with sulfur dioxide and water; other instances are sulfur and toluene, sulfur and xylene.[3] Now that the conditions have been clearly defined under which such a system occurs there will be no difficulty in increasing the number of illustrations at pleasure. If a solid melt under a liquid with formation of a second liquid phase, there will be, near the temperature at which this takes place, a solution saturated with respect to a solid and having a higher vapor pressure than either pure component or two liquid phases with an intermediate vapor pressure. In other words, of two cases, both of which have been considered abnormal and even impossible, one always occurs.

A third possibility in the way of two liquid phases and vapor is that the vapor pressure of this system shall be less than that of either component. This would be represented by a curve of the same general form as LMNO. No example illustrating this has been found[4]; but an approximation to it has been found in the system, hydrobromic acid and water, which is represented in the diagram by PQRST. PQR is the curve for the unsaturated solution of hydrobromic acid in water. At small concentrations the solution has a lower vapor pressure than either component; but with increasing concentration the vapor pressure of the divariant system passes through a minimum, represented by Q, and increases rapidly to the point R where the second liquid phase appears with and the system passes along RS. ST is the curve for the solution of water in hydrobromic acid. I have been unable to find any data in regard to

---

[1] Ostwald, Lehrbuch I, 643.
[2] Recueil Trav. Pays-Bas **3**, 31, (1884); Zeit. phys. Chem. **8**, 526, (1891).
[3] Bancroft, Jour. Phys. Chem. **1**, No. 3 (1896); Haywood, Ibid. **1**, No. 4 (1897).
[4] Konowalow considers it unrealizable. Wied. Ann. **14**, 221, (1881).

the direction of this curve; but it probably ascends if one may judge from the behavior of sulfur dioxide and water.

Since the two liquid phases in equilibrium have very different compositions, it may be asked what is the relation between the vapor pressure of each of the liquid phases and of the system. Konowalow[1] showed experimentally and theoretically that the two phases must give off the same vapor and that this was the vapor in equilibrium with the two solution phases. Ostwald[2] reached the same conclusion in much the same way. I quote his reasoning: "Suppose that in a hollow ring (Fig. 19) A is a satur-

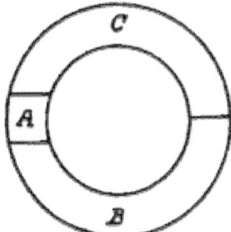

FIG. 19.

ated solution of water in ether, B of ether in water, and C is the vapor space. If the vapor in equilibrium with A had a different pressure or composition from the vapor in equilibrium with B, there would be a continuous distillation or diffusion from one side to the other and equilibrium would never be reached since the original condition would always be restored by diffusion through the surface between the solutions. We should thus have a perpetual motion machine, which is impossible." This argument is not quite sound.[3] If the partial pressure of each solute be higher than the partial pressure of the same component in the solution in which it is solvent, Ostwald's description of what will take place is accurate. The distillation from a place of high pressure to a place of low pressure will produce unsaturated solutions which cannot exist in contact and this assumption is, therefore, untenable. The case is quite different if

[1] Wied. Ann. **14**, 220, 224, (1881).
[2] Lehrbuch, I, 644.
[3] Bancroft, Phys. Rev. **3**, 415, (1896).

the partial pressure of each component be assumed to be higher in the vapor above the solution in which it is solvent than in the vapor above the solution in which it is solute or, to take a specific case, if there be more ether in the vapor given off by A than in the vapor given off by B, and more water vapor given off by B than by A. Ether will distill from A to B and water from B to A. The water condensing at A will sink through the less dense ether layer; the ether which condenses at B will become saturated with water forming an infinitely thin film of a solution of water in ether on top of the aqueous solution B. Both free surfaces having the same composition, there is no possibility of further distillation if the influence due to gravity be neglected. It has been pointed out to me by Professor Trevor that the two solutions of water in ether, not being at the same level, will have different vapor pressures owing to the influence of gravity and that a final equilibrium will not be reached until the system has become symmetrical. This experiment has been tried in his laboratory and the prediction verified.

It is thus possible for two liquid phases to be in equilibrium without assuming that each sends off the same vapor. It is possible that both the pressures and the compositions of the two vapors may differ, and consequently that the vapor in equilibrium with the two liquid phases may differ both in pressure and composition from the vapor in equilibrium with either of the saturated solutions. Under these circumstances the horizontal part of the isothermal curve would not meet the two side curves but would be connected with them by two practically vertical lines representing the state of things, while the second solution phase is represented in such small quantity as not to have the properties of matter in mass.[1] Whether such cases actually occur is open to doubt; but they are not theoretically impossible on the ground of violating the law of the conservation of energy. If one admits the possibility of two liquids being absolutely non-miscible, that would at once furnish such a case and might be exemplified by benzene and mercury for instance. If one assumes that all substances, liquid or solid, are miscible to a certain extent,[2] the

---

[1] Gibbs, Trans. Conn. Acad. **3**, 129 (1876).
[2] Nernst, Theor. Chem. 393.

question can only be settled by careful measurements. If the vapor given off by the two phases are identical, this is a proof that the change of partial pressure is not the same function of the concentration for the solvent and the solute.[1]

Instead of measuring pressures at constant temperature, it is easier to measure temperatures at the constant atmospheric pressure and to plot the change of the boiling point with the concentration. This will give a curve of the same general form as the pressure-concentration curve, but reversed, because a low vapor pressure implies a high boiling point and *vice versa*. The two forms of boiling point curves for partially miscible liquids which have been realized experimentally are given in Fig. 20. They refer to the same pairs of liquids as the similarly lettered curves in Fig. 18. Thus ABCD is the typical boiling point curve for most pairs of liquids.[2] As will be seen from the diagram, the monovariant system, two liquid phases and vapor has a

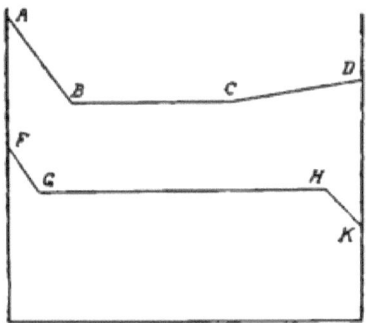

FIG. 20.

lower boiling point than either of the pure components. When the monovariant system, two liquid phases and vapor, has a vapor pressure lying between those of the single components, the boiling point of the system will lie between the boiling points of the pure substances. The only instances of two liquids boiling in an open flask at temperatures between the boiling points of the components are sulfur and toluene, sulfur and xylene. The boiling point curves for these systems will have the general forms represented by FGHK.

[1] Bancroft, Phys. Rev. **3**, 203 (1895).
[2] Cf. Konowalow, Wied. Ann. **14**, 142 (1881).

While these two types show the same general behavior, a constant boiling point, so long as the monovariant system is present, they behave very differently when it comes to the divariant system left behind. With two liquids of the first type as, for instance, isobutyl alcohol and water, either of the liquid phases may disappear on distillation, depending on the relative quantity of the two phases. The single liquid phase left in the flask may therefore be a solution of isobutyl alcohol in water, or a solution of water in isobutyl alcohol. With sulfur and xylene, on the other hand, it will always be the phase, solution of sulfur in xylene, which disappears on distillation, and the liquid in the flask will always be a solution of xylene in sulfur, entirely irrespective of the relative amounts of the two liquid phases in the original mixture. This difference in behavior is shown in the diagram; both side curves ascend in systems of the first type, only one in liquids of the second type. Only a mixture with a higher boiling point can be left behind in the flask, because the temperature never sinks during a distillation.[1] Since all solutions of sulfur in xylene boil at a lower temperature than the monovariant system, they can never occur as residues.

In the diagram for naphthalene and water, Fig. 17, the curves BC and ED have only been followed a little way; but it will be profitable to consider what will be the effect of increasing temperature on a system composed of two liquid phases and vapor. The direction of the curves for the concentrations of the two phases can be foretold from the Theorem of Le Chatelier. If the solute dissolves with absorption of heat, its concentration will increase with rising temperature, and *vice versa*. While it is distinctly exceptional for a solid to dissolve with absorption of heat at ordinary temperatures, this is by no means uncommon in the case of a liquid solute. There are three possibilities in a system composed of two liquid phases and vapor. With rising temperature there may be increasing solubility in both phases, or increasing solubility in one phase and decreasing solubility in the other, or decreasing solubility in both phases. These three cases have been realized experimentally by Alexejew.[2] Phenol and

---

[1] An empirical generalization.
[2] Wied. Ann. **28,** 305 (1886).

water is an example of the first case, the solubility of each in the other increasing with increasing temperature.[1] Sulfur and toluene is another instance. The solubility of water in isobutyl alcohol increases with the temperature, that of isobutyl alcohol in water decreases at all temperatures below 65°. Other examples are mixtures of water with ether[2] or esters.[3] In all these instances it is the solubility in water which decreases with increasing temperature. Between the temperatures of 0° and 65°, both the solubility of water in secondary butyl alcohol and that of secondary butyl alcohol in water decreases with increasing temperature. A yet more striking example of this class is to be found in mixtures of diethylamine and water, investigated by Guthrie.[4] Alexejew found that on increasing the temperature sufficiently all pairs of liquids finally come under the first heading, with solubilities increasing with increasing temperature. Under these circumstances the compositions of the two liquid phases will approach each other until at some temperature they become equal, and there is only one liquid phase. Above this temperature the liquids are miscible in all proportions, or "consolute." The temperature at which this takes place is known as the consolute temperature.[5] The same result has been reached in a different way by Masson,[6] starting from the experimental fact that gases are miscible in all proportions. From this it follows that two liquids must become consolute at the critical temperature of the mixture, and it is probable that many pairs will do so at a much lower temperature. In Fig. 21 are the concentration-temperature curves for some of the systems studied by Alexejew, showing the different typical cases. There is added also the curve for diethylamine and water, because the increase of solubility in the two phases with decreasing temperature is so great that the liquids are consolute below a given tempera-

---

[1] If the assumption is made that water is the solvent in both liquid layers, the concentration of phenol in the phase containing an excess of that substance should increase with rising temperature, which is not the case.

[2] Nernst, Theor. Chem. 390. [3] Bancroft, Phys. Rev. 3, 132 (1895).

[4] Phil. Mag. (5), 18, 500 (1886).

[5] Bancroft, Jour. Phys. Chem. 1, No. 3 (1896).

[6] Zeit. phys. Chem. 7, 500 (1891).

ture.[1] Unless decomposition occurred it would be possible to find a second temperature above which the two liquids would be miscible in all proportions.

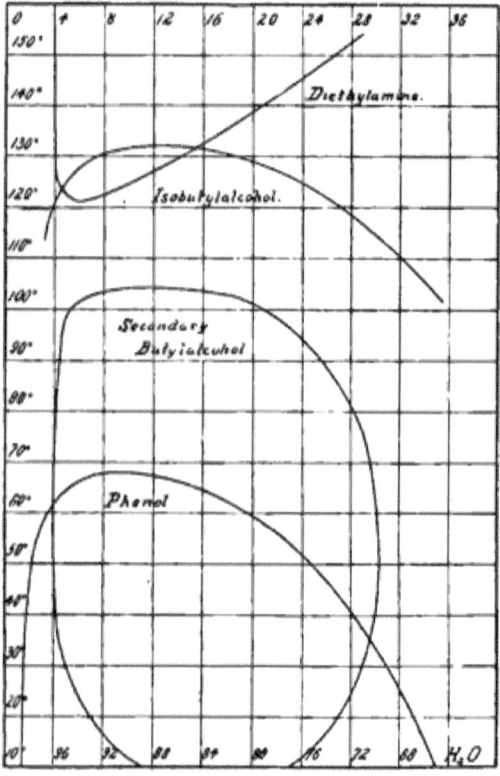

FIG. 21.

In the system made up of naphthalene and water, the consolute temperature is so far above the melting point of naphthalene that it does not appear on the diagram, Fig. 17, and the two liquid phases can be in equilibrium over a wide range of temperatures. With phenol and water this is no longer the case, and this concentration-temperature diagram for this equilibrium is represented in Fig. 22. AB

---

[1] The same phenomenon occurs with ethylamine and with triethylamine.

and HF are the fusion curves for phenol and ice respectively.[1] MBL and NDL are the solubility curves for the two liquid phases, water in phenol and phenol in water, the portions BM DN representing states of labile equilibrium unstable in respect to solid phenol. L is the consolute temperature, 67°–68°, and DF is the solubility curve for solid phenol. Sometimes it occurs that the consolute temperature is

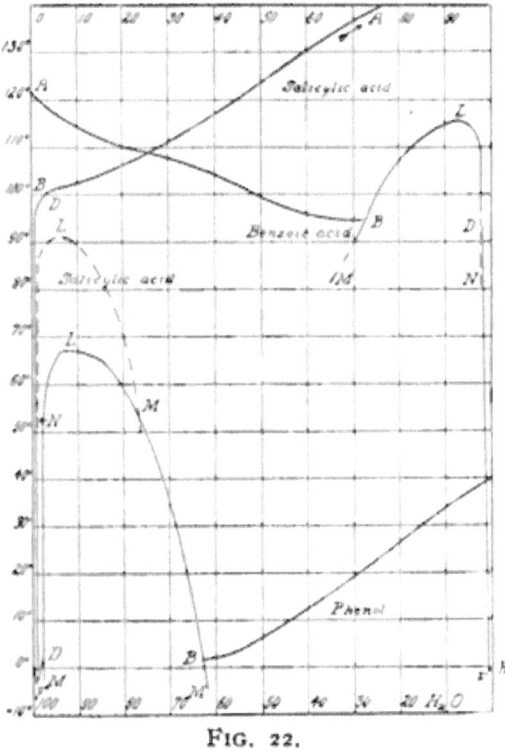

FIG. 22.

lower than the melting point of the less fusible component, as in the case of benzoic acid[2] and water, Fig. 22, while with salicylic acid

---

[1] The letter H has been omitted to avoid crowding. It belongs at the point marked 0°. For the letter M in the lower left-hand square of the diagram, read N.

[2] To avoid confusion this curve has been reversed, one hundred units of water being at the right. The wavy form of the curve AB is probably due to experimental error.

and water, Fig 22, the whole curve MLN is instable.[1] Here AB is unquestionably a fusion curve, DF equally unquestionably a solubility curve, and at some point on the curve ABDF there must therefore be a change from one to the other; but at present we are not able to determine that point exactly, nor to say anything in regard to the curve BD if B and D do not coincide. It is very much to be desired that these measurements of Alexejew should be repeated with great care, as it seems not impossible that the curve BD is really a straight line parallel to the temperature axis, and that there is a sudden change of direction both at B and at D. A still more remarkable series of curves is furnished by triethylamine and water in Fig. 23. The mixtures of the two liquids become turbid on warming,

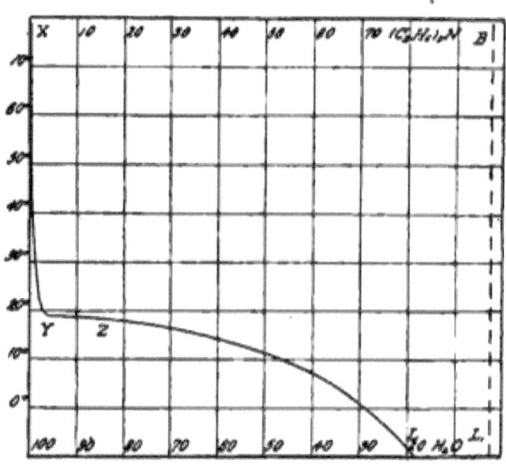

FIG. 23.

separating into two liquid layers. The curve XYZL represents the temperatures at which this takes place for solutions containing from five to ninety-eight per cent. of water in grams. In the diagram one hundred and one grams of triethylamine are taken as equivalent to eighteen grams of water. No points on the curve were determined beyond L, but the statement that a solution containing one per cent.

---

[1] Roozeboom has observed a similar phenomenon with water and a salt of trinitrophenylmethylnitramine. Recueil Trav. Pays-Bas, **8**, 263 (1889).

of water did not become turbid even **at 200°**, shows that the curve L_1B must represent the behavior of **solutions** containing very little water. The portion of the curve YZ was believed by Guthrie to be actually horizontal; but what this signifies is entirely unknown.[1]

In the **nonvariant system**, solid, **two** solutions **and vapor, the solid phase is not always one of** the pure components; **but may be a compound. A most** excellent example of this form **of equilibrium has been found by** Roozeboom[2] in the system, composed **of sulfur dioxide and water.** The solid phases which occur are **ice and the hydrate $SO_2,7H_2O$.** The symbols $H_2O \rightleftharpoons xSO_2$ and $SO_2 \rightleftharpoons yH_2O$ de**note a** solution of x reacting weights of **SO_2 in** one reacting **weight** of water, and a solution of y reacting weights of **water in one of** SO_2, water being solvent in the first case **and sulfur dioxide in the** second. The pressure-temperature diagram **for this system is shown** in Fig. 24. At B **there exists the nonvariant system, ice, hydrate,** $H_2O \rightleftharpoons xSO_2$ and **vapor. The temperature is** $-2.6°$ **and the pres**sure 21.1 cm. **of mercury. BF, BC, BZ and BL are the boundary** curves **for** ice, $H_2O \rightleftharpoons xSO_2$, **and vapor; ice, hydrate and vapor; ice, hydrate and** $H_2O \rightleftharpoons xSO_2$; **hydrate, $H_2O \rightleftharpoons xSO_2$ and vapor respec**tively, while BA is the **labile prolongation of LB, instable with re**spect to ice. **Passing along the curve BL, there is formed at L** a new nonvariant system composed **of hydrate, $H_2O \rightleftharpoons xSO_2$,** $SO_2 \rightleftharpoons yH_2O$ and vapor. The temperature **at which this system can** exist is $12.1°$ and the pressure 177.3 cm. **of mercury. The curves** LE, **LX and** LD **represent the monovariant systems, $H_2O \rightleftharpoons xSO_2$,** $SO_2 \rightleftharpoons yH_2O$ and vapor; **hydrate, $H_2O \rightleftharpoons xSO_2$ and $SO_2 \rightleftharpoons yH_2O$; hydrate, $SO_2 \rightleftharpoons 7H_2O$ and vapor. There is no nonvariant system** possible at the point where **the curve LD cuts BZ because the two curves have only one phase in common, the hydrate $SO_2,7H_2O$.**[3] LD will terminate at **some** low temperature **owing to the formation of** solid sulfur dioxide, forming the **nonvariant system, SO_2, $SO_2,7H_2O$,** $SO_2 \rightleftharpoons yH_2O$ and vapor; but **no attempt has yet been made to realize**

---

[1] Pickering has made **a careful study of the freezing points of amines and** water. Jour. Chem. Soc. **63**, 141 (1893).

[2] Recueil Trav. Pays-Bas, **3**, 29 (1884); Zeit. phys. Chem. **2**, 450 (1888).

[3] Private letter from Professor Roozeboom.

this equilibrium. The curve EL, if continued, will lie below LD, theoretically;[1] but the difference in direction falls within the limits of experimental error and, therefore, does not show in the diagram.

Since the systems in equilibrium with solid sulfur dioxide were not investigated, there is but one fusion curve, FB, the other curves

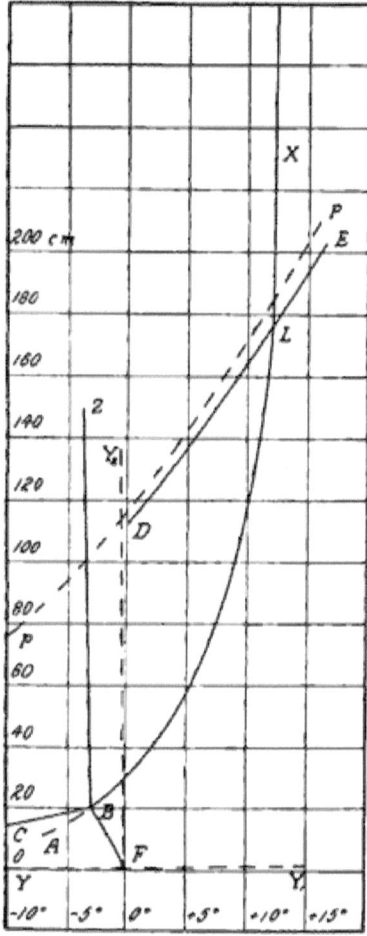

FIG. 24.

---

[1] Nernst's diagram is faulty in this point. Theor. Chem. 488.

for systems containing liquid phases being solubility curves. Along LB water is solvent and along LD sulfur dioxide. Along LX and LE sulfur dioxide is solvent in one of the two liquid phases and solute in the other. It will be seen that the equilibrium between sulfur dioxide and water differs in one respect from anything that has yet been considered. The solvent is not the same in the two solutions which may exist in equilibrium with the solid hydrate $SO_2 \cdot 7H_2O$, while water is solvent in both the solutions saturated with respect to $Fe_2Cl_6 \cdot 12H_2O$, to take merely one instance. It might seem at first as if this difference were connected with the facts that the solid hydrate, $SO_2 \cdot 7H_2O$, does not have a true melting point, and that it can exist in equilibrium with two liquid phases simultaneously; but neither of these things is essential. There are no theoretical reasons known why the whole of the curve LE should not represent a labile equilibrium, as was found to be the case experimentally with salicylic acid and water. Under these circumstances the hydrate would have a true melting point, and we could not say with our present knowledge whether the solvent changed or not. This point can not be settled definitely in all cases until we are able to say at what concentration a mixture of two consolute liquids changes from a solution of one in the other to a solution of the second in the first.

In order to define the fields in which the divariant systems can exist, it will be necessary to add, in Fig. 24, parts of the pressure-temperature diagrams for each of the pure components. The dotted line pp represents the equilibrium between liquid and gaseous sulfur dioxide; YF between ice and vapor; FY_1 between water and water vapor; FY_2 between ice and water. Ice and hydrate exist in the field ZBC; ice and $H_2O \rightleftharpoons xSO_2$ in $ZBFY_2$; ice and vapor in CBFY; hydrate and $H_2O \rightleftharpoons xSO_2$ in ZBLX; hydrate and vapor in CBLD; $H_2O \rightleftharpoons xSO_2$ and vapor in Y,FBLE; hydrate and $SO_2 \rightleftharpoons yH_2O$ in XLD; $H_2O \rightleftharpoons xSO_2$ and $SO_2 \rightleftharpoons yH_2O$ in XLE, while $SO_2 \rightleftharpoons yH_2O$ can be in stable equilibrium with vapor only in the space between the lines DLE and pp.[1] Increase of pressure causes any system represented by a point on LB to pass into hydrate and vapor, if the quantity of vapor in the monovariant system be sufficiently large relative-

---

[1] Roozeboom, Recueil Trav. Pays-Bas, **6**, 319 (1887).

ly to the solution; into hydrate and solution of sulfur dioxide in water if the contrary is true. A further increase of pressure produces the monovariant system represented by LD on the first assumption, while, in the second case, the same change will form the monovariant system, hydrate and two liquid phases, if accompanied by a rise of temperature. If the monovariant system, hydrate, water in sulfur dioxide, and vapor, be compressed, either the hydrate or the vapor will disappear, depending on the relative quantities of the two. Starting from any point on LF, the two phases which decrease with increasing pressure are the solution of sulfur dioxide in water and the vapor; again, it is a question of relative amounts whether one or the other of the two possible divariant systems be formed. This will be clear if it be kept in mind that the vapor contains both sulfur dioxide and water, and that the ratio of sulfur dioxide to water in the vapor is greater than the ratio of the two components in the solid hydrate or in the solution of sulfur dioxide in water and less than the ratio in the solution of water in sulfur dioxide. As the concentrations of the various solutions were not all determined, it is impossible to make the concentration-temperature diagram for this particular case; but in Fig. 25 is given the general form for two

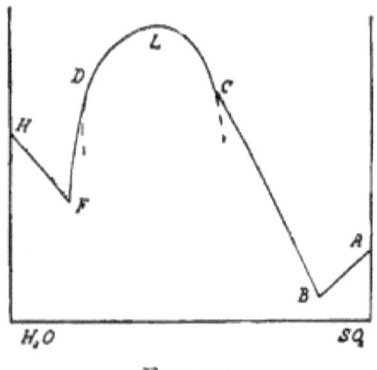

FIG. 25.

components which form one compound and two liquid phases. AB is the fusion curve for SO_2 in presence of water. At B the hydrate crystallizes out forming the nonvariant system, solid SO_2, $SO_2 \cdot 7H_2O$, solution of water in SO_2, and vapor. Along BC we have the solubility curve of the hydrate, SO_2 being the solvent. CL and DL are the two liquid phases and L the consolute temperature, SO_2 being solvent along CL and water along DL. At the temperature represented by C and D there exists the nonvariant system, hydrate, two solutions and vapor. DF is the solubility curve for the hydrate, water being solvent, while HF is the fusion curve for ice in presence of SO_2. At F there exists the nonvariant system, ice, hydrate, solution of SO_2 in water, and vapor. This is on the assumption that the critical temperature of sulfur dioxide is higher than that of the hypothetical point L. Tables XIX-XX give Roozeboom's data for sulfur dioxide and water, sixty-four grams of the former being equivalent to eighteen of the latter. The pressures are in centimeters of mercury except for the system, hydrate and two liquid phases, where the values are given in atmospheres.

Table XIX

Nonvariant Systems	Temp.	Pressure
Ice, hydrate $H_2O \rightleftarrows 0.024 SO_2$, vapor	$-2.6°$	21.15
Hydrate $H_2O \rightleftarrows 0.087 SO_2$, $SO_2 \rightleftarrows yH_2O$, vapor	12.1	177.3

Table XX

Temp.	Pressure	Temp.	Pressure	Temp.	Pressure
Ice, hydrate vapor		Hydrate, $H_2O \rightleftarrows xSO_2$ vapor		$H_2O \rightleftarrows xSO_2$, $SO_2 \rightleftarrows yH_2O$ vapor	
$-9.°$	15.0	$-6.°$	13.7	$12.1°$	177.3
$-8.$	16.0	$-4.$	17.65	13.0	182.3
$-6.$	17.7	$-3.$	20.1	Hydrate, $H_2O \rightleftarrows xSO_2$	
$-4.$	19.35	-2.6	21.15	$SO_2 \rightleftarrows yH_2O$	
$-3.$	20.65	$-2.$	23.0	12.1	0.233
-2.6	21.15	$-1.$	26.2	12.9	20.
Hydrate, $SO_2 \rightleftarrows yH_2O$ vapor		0.0	29.7	13.6	40.
		2.8	43.2	14.2	60.
$+0.1°$	113.1	4.45	51.9	14.8	80.
3.05	127.0	8.00	66.6	15.3	100.
6.05	141.9	8.40	92.6	15.8	125.
9.05	158.2	10.00	117.7	16.2	150.
11.0	170.1	11.30	150.3	16.5	175.
11.9	176.2	11.75	166.6	16.8	200.
12.1	177.3	12.10	177.3	17.1	225.

So far as is known, sulfur dioxide and water form but one compound, $SO_2 7H_2O$, and this does not have a true melting point.[1] Hydrobromic acid and water form four solid compounds, $HBr4H_2O$, $HBr3H_2O$, $HBr2H_2O$ and $HBrH_2O$, all except the last having a true melting point. Fig. 26 is the pressure-temperature diagram for part of this system,[2] drawn to scale. The pressures are given in atmos-

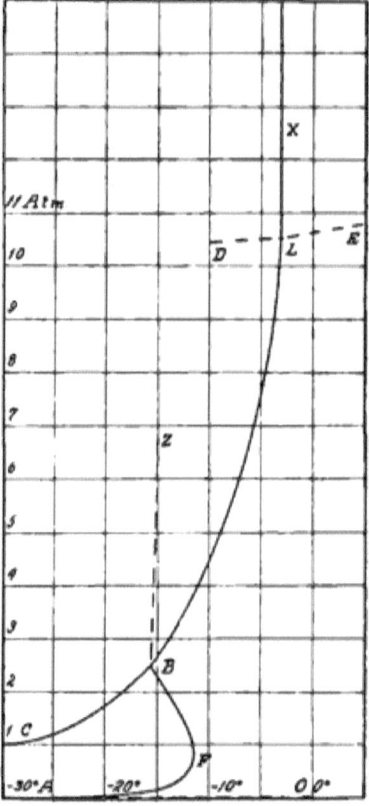

FIG. 26.

[1] Roozeboom has also studied the equilibrium between chlorine and water. Recueil Trav. Pays-Bas, **3**, 59 (1884); **4**, 69 (1885); between bromine and water. Ibid. **3**, 73 (1884); **4**, 71 (1885).

[2] Roozeboom, Ibid. **4**, 108, 331 (1884); **5**, 323, 351, 353 (1885).

pheres. OA is the fusion curve for ice in presence of hydrobromic acid. The line is dotted because this equilibrium was not studied experimentally by Roozeboom and does not appear in his diagram.[1] Along AFB the dihydrate exists in equilibrium with solution and vapor. At F the solution has the same composition as the hydrate. The temperature of this melting point is -11.30 and the pressure 52.5 cm. of mercury. Along FB the hydrate is in equilibrium with a solution containing more hydrobromic acid than itself. Water is still the solvent so this is analogous to the hydrates of ferric chlorides. At B there exists the nonvariant system, $HBr2H_2O$, $HBr.H_2O$, $H_2O \rightleftharpoons rHBr$ and vapor, the temperature being $-15.5°$ and the pressure two and one-half atmospheres. BC represents the conditions under which the dihydrate and monohydrate can be in equilibrium with vapor while BL is the curve for monohydrate, solution and vapor. The co-ordinates of the point L are $-3.3°$ and ten and one-half atmospheres. The nonvariant system existing under these conditions is $HBrH_2O$, $H_2O \rightleftharpoons rHBr$, $HBr \rightleftharpoons rH_2O$ and vapor. The curves radiating from L have the same lettering and significance as those in Fig. 24 if one substitutes hydrobromic acid for sulfur dioxide and remembers that the solid phase has the composition $HBrH_2O$. The solution along LB contains less hydrobromic acid than the crystals and if cooled at constant volume will solidify completely at B with formation of monohydrate, dihydrate and vapor. Roozeboom performed no experiments below $-30°$, so that he studied only the behavior of the two compounds, $HBr2H_2O$ and $HBrH_2O$. Pickering[2] has shown that there is a compound $HBr3H_2O$ with a melting point at $-48°$ and a compound $HBr4H_2O$ with a melting point at $-55.8°$. In Fig. 26 the lines OA and BFA are drawn and lettered as if they came together at A. This is obviously inaccurate but the lines could not be kept separate unless a very different scale were used. In reality the line BFA terminates at about $-48°$ with the appearance of the trihydrate and the formation of the monovariant system, $HBr3H_2O$, solution and vapor. At about $-57°$ there can exist the nonvariant system, $HBr4H_2O$,

---

[1] Zeit. phys. Chem. **2**, 454 (1888).
[2] Phil. Mag. (5) **36**, 111 (1893).

$HBr3H_2O$, solution and vapor. The tetrahydrate melts at $-55.8°$. The dotted curve OA has been followed as far as $-73°$ and the precipitate was still ice. It is not known at what temperature the next nonvariant system occurs nor whether the new phase is tetrahydrate or some other compound. The systems, hydriodic acid and water, hydrochloric acid and water, are very similar to the one just discussed and bring out no new points.[1] In tables XXI-XXIII are Roozeboom's data for hydrobromic acid and water. x_1 denotes units of hydrobromic acid per unit of water; x_2 units of water per unit of hydrobromic acid; x_3 units of hydrobromic acid in one hundred units of solution. Eighty-one grams of hydrobromic acid are equivalent to eight grams of water.

TABLE XXI

Nonvariant Systems	Temp.	Pressure
$HBr2H_2O$, $HBrH_2O$, $H_2O \rightleftharpoons 0.61HBr$, vapor	$-15.5°$	195 cm.
$HBrH_2O$, $H_2O \rightleftharpoons 0.83HBr$, $HBr \rightleftharpoons yH_2O$, vapor	-3.3	10½ Atm.
Melting point of $HBr2H_2O$	-11.3	525 cm.

TABLE XXII

Temp.	Pressure	x_1	x_2	x_3	Temp.	Pressure	x_1	x_2	x_3
$HBr2H_2O$, solution and vapor.					$HBr2H_2O$, solution and vapor				
$-25.°$	0.1 cm.	0.39	2.56	28.0	$-11.5°$	1 Atm.	0.52	1.91	34.3
-21.8	1.				-12.0	1¼	0.54	1.85	35.1
-18.9	3.	0.42	2.38	29.6	-12.6	1½	0.55	1.81	35.6
-16.8	6.				-13.3	1¾			
-14.6	12.	0.45	2.24	30.8	-14.0	2	0.59	1.70	37.0
-13.0	22.				-14.8	2¼			
-12.4	28.				-15.5	2½	0.61	1.63	38.0
-12.0	34.	0.48	2.10	32.3					
-11.6	44.								
-11.3	52.5	0.50	2.00	33.3					

---

[1] Roozeboom, Recueil Trav. Pays-Bas, **3**, 84 (1884); Zeit. phys. Chem. **2**, 459 (1888); Pickering, Phil. Mag. (5) **36**, 111 (1893).

Temp.	Pressure	x_1	x_2	x_3
\multicolumn{5}{c}{$HBrH_2O$, solution and vapor.}				
−15.5°	2½ Atm.	0.61	1.64	37.9
−14.8	2¾			
−14.0	3	0.62	1.61	38.3
−11.0	4	0.65	1.55	39.3
− 8.7	5	0.67	1.50	40.0
− 7.2	6	0.69	1.45	40.9
− 5.8	7	0.72	1.38	42.0
− 4.7	8	0.76	1.32	43.1
− 4.0	9			
− 3.3	10½	0.83	1.20	45.5

TABLE XXIII

Temp.	Pressure	Temp.	Pressure	Temp.	Pressure	Temp.	Pressure
\multicolumn{4}{c}{$HBr2H_2O$, $HBrH_2O$ and vapor}	\multicolumn{4}{c}{$HBrH_2O$, $H_2O \rightleftarrows xHBr$, $HBr \rightleftarrows yH_2O$}							
−28.5°	76 cm.	−20°	131 cm.	−3.3°	10½ Atm	−1.6°	100 Atm.
−26.	85	−18	156	−2.9	25	−0.9	150
−24.	96	−16	184	−2.4	50	−0.3	200
−22.	111	−15.5	195	−2.0	75	+0.3	250

CHAPTER VIII

CONSOLUTE LIQUIDS

Since a fusion curve ends at its intersection with the solubility curve, it is clear that the greater the solubility the farther the fusion curve can be followed and that this may even be carried to such an extent that the fusion curve for one component may meet the fusion curve for the other component. This actually occurs in a great many instances and has often been looked upon as the typical case from which all equilibria could be derived by adding the necessary limitations. It is more rational to treat it as a subhead of the normal case, exemplified in Figs. 17, 22, in which the solubility curves have disappeared. As an example we will take the equilibrium between naphthalene and phenanthrene studied by Miolati,[1] Fig. 27.

[1] Zeit. phys. Chem. **9**, 649 (1892).
[2] Phil. Mag. (5) **17**, 462 (1884).

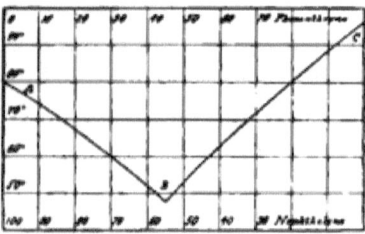

FIG. 27.

Addition of phenanthrene lowers the freezing point of naphthalene and the temperatures and concentrations at which the monovariant system, naphthalene, solution and vapor, can exist are shown by the curve AB. Starting with phenanthrene and adding naphthalene there is a lowering of the freezing point and the curve CB represents the conditions of equilibrium for the system, phenanthrene, solution and vapor. At B, 48°, there exists the nonvariant system, naphthalene, phenanthrene, solution and vapor. At this point the solution solidifies without change of temperature. The mixture with the lowest freezing point was called by Guthrie the eutectic alloy and he looked upon the cryohydric mixture as a special case of the same thing. There is one important difference between the two cases. Salt and ice appear at the intersection of a solubility and a fusion curve, naphthalene and phenanthrene at the intersection of two fusion curves. It is desirable to distinguish between these two cases in the following way. The temperature at which a binary solution in equilibrium with vapor passes completely into two solid phases is called the cryohydric temperature, if a solubility and a fusion curve meet at this point and the eutectic temperature if the intersection is between two fusion curves. This does not include the third case in which a solution can solidify without change of temperature when two solubility curves meet, and the solution has a composition between those of the two solid phases. This is realized when $Fe_2Cl_6 12H_2O$, $Fe_2Cl_6 7H_2O$, solution and vapor are in equilibrium, but there seems to be no need of a special name for this temperature.

In the field ABC there exists the divariant system, solution and vapor. At all temperatures higher than the melting point of the less fusible component, there can be no solid phase if there are no compounds possible, and the solution will be a mixture of two consolute liquids. If the two components form compounds or solid solutions, it is merely necessary to raise the temperature above the melting point of the least fusible solid phase in order to reach the same state of things. The divariant system, solution and vapor, is also made up of two consolute liquids when the temperature of the experiment is higher than that at which two liquid phases can coexist with vapor. Strictly speaking, there are two classes of consolute liquids, those which never form two liquid phases under any circumstances and those which sometimes do. Our knowledge of the subject is, unfortunately, so imperfect that it is impossible from a study of the vapor pressures of mixtures of consolute liquids to predict what new phase will separate under given conditions. The different ways in which the pressure changes with the concentration at constant temperature are shown in the concentration-pressure diagram, Fig. 28. Four styles of curves have been found experimentally, having a maximum, a minimum, both a maximum and a minimum and neither a maximum nor minimum value. In systems represented by AA, addition of either component to the other produces an increase in the vapor pressure, and there will therefore be some concentration for which the vapor pressure has a maximum value higher than that of either of the pure components. Instances of this are to be found in mixtures of propyl alcohol[1] or butyric acid with water; in mixtures of carbon bisulfide with ethyl alcohol or ethyl acetate;[2] in carbon tetrachloride and methyl alcohol.[3] In systems represented by BB,

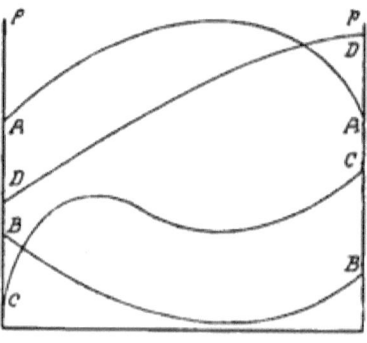

FIG. 28.

addition of either compound to the other causes a lowering of the vapor pressure so that there will be some concentration of the liquid phase in equilibrium with a vapor having a minimum pressure lower than that of either of the components. Mixtures of water with formic, nitric, and the haloid acids come under this head.[4] There are no instances known of an isothermal pressure curve having a maximum and a minimum value, the one being higher and the other lower than that of either component, though this is a perfectly conceivable case and might, pehaps, be realized if some one were to study mixtures of one of the haloid acids and water at a suitable temperature. There are two instances known where the isothermal pressure

---

[1] Konowalow, Wied. Ann. **14**, 50 (1881).
[2] Brown, Jour. Chem. Soc. **39**, 529 (1881).
[3] Thorpe, Ibid. **35**, 544 (1879).
[4] Konowalow, Wied. Ann. **14**, 51 (1881).

curve has the form represented by CC, with both a maximum and a minimum value, lying between those for the pure components. These two cases are propionic acid and water[1] at 64°, benzene and carbon tetrachloride[2] at 34.8°. It is by no means certain that this form of curve really exists. Vapor pressure measurements are not easy to make accurately, and it may well be that the wavy nature of the curve is due to experimental error. With propionic acid and water this is the more probable conclusion, since the difference between the maximum and the minimum is small and the data for constructing the curve were obtained by interpolation and not by direct measurement. With benzene and carbon tetrachloride the matter is a little different. It is necessary to assume an error of some two centimeters in a total of seventeen in order to account for the phenomenon. This means an error of at least ten per cent., and would detract greatly from the importance of Linebarger's work if established. It is much to be regretted that the unusual nature of the results was not recognized and especial pains taken to verify or disprove them. The fourth class of curve is that represented by DD, having neither maximum nor minimum. The curve may be either convex or concave, and is typical of most pairs of consolute liquids. It is not known what is the relation of this curve to the others. If two consolute liquids have the same vapor pressure at a given temperature, some mixture of the two must have either a maximum or a minimum vapor pressure at that temperature unless all concentrations have the same value. If at other temperatures the isothermal curve has neither maximum nor minimum it is possible to pass by change of temperature from a system behaving in one way to a system behaving in another. This would prevent any deductions from the form of the pressure curve at one temperature to that at another. The alternative is that all consolute liquids with intersecting vapor pressure curves[3] form solutions with a maximum or minimum value at some concentration. No experimental work on this subject has been done.

---

[1] Konowalow, *l. c.* 45.
[2] Linebarger, Jour. Am. Chem. Soc. **17,** 690 (1895).
[3] Guye, Zeit. phys. Chem. **14,** 570 (1894).

In order to treat the subject of fractional distillation, it is necessary to consider the boiling points of mixtures of consolute liquids. In the concentration-temperature diagram, Fig. 29, are represented the three important types of boiling point curves. The systems which have a maximum vapor pressure have a minimum boiling point and *vice versa*, while systems with the vapor pressures of all mixtures lying between the values for the pure components will have intermediate boiling points. Although the quantitative data in regard to the composition of the vapor in equilibrium with a given solution are few and far between, there is little difficulty in determining them qualitatively, and these compositions are given schematically by the dotted lines in the diagram.

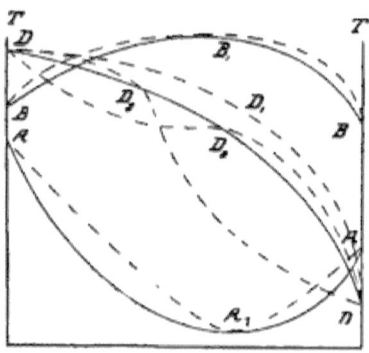

FIG. 29.

The composition of the vapor at any temperature is given by the point at which the horizontal line for that temperature meets the dotted curve. At the temperature minimum and maximum, A_1 and B_1, the vapor has the same percentage composition as the liquid and these solutions will therefore distill without change of temperature. This has been confirmed experimentally, the mixture of propyl alcohol and water with the lowest boiling point and that of formic acid and water which has the highest boiling point, behaving like pure liquids. All other mixtures in so far as they are represented in the diagram have the two components present in different proportions in the vapor and the liquid phases. For convenience it will be assumed that the left side of the diagram represents one hundred per cent.

water and the right side **one hundred per cent. of the other component**. In the equilibrium between propyl **alcohol and water represented by AA₁A all the solutions to the left of A₁ contain more water than the vapor, while all to the right contain more propyl alcohol than** the vapor. Partial distillation yields **a distillate richer in propyl** alcohol than the **residue in** the **first case and one richer in water in the second case.** By continued or fractional distillation of a mixture of propyl alcohol and water, **there is** obtained **in the distillate the mixture with the** lowest boiling point **and** composition **A₁ which cannot be purified** further in this way. In the distilling **flask there is left behind** pure water if the original solution **contained more** water than the mixture with constant boiling point and **pure propyl** alcohol if the contrary was the case. With liquids like **formic acid** and water, BB₁B, the phenomena are **reversed.** Solutions to the left of B₁ contain less water than the vapor; **solutions to the right less** formic acid. By fractional distillation the **first set give pure water, the** second pure formic acid in the distillate, **while both leave behind in the flask the solution of formic acid and water with the highest** boiling point. This cannot **be separated into its components by** further fractioning, because, as **has been pointed out, the solution** and vapor have the same composition. In **Table XXIV are given**

TABLE XXIV

H₂O	HCl	100°	−88°	110°	20.2
H₂O	HBr	100	−64	116	47.8
H₂O	HI	100	−34	127	57
H₂O	HCOOH	100	+99.9	107.1	77.5
H₂O	HNO₃	100	86	120.5	68.0
HCl	(CH₃)₂O	−83	−21	−2	61
H₂O	C₃H₇OH	100	+97.4	+97.4	77
CCl₄	CH₃OH	76.6	65.2	65.2	21.9

the compositions and boiling **points of some mixtures** corresponding to the points A₁ and B₁. In the first and second columns are the formulas representing the two components; **in the third and fourth columns** the boiling points of **the** first **and second** components respectively; in the fifth, the boiling point of the mixture with constant composition, and in the sixth the composition of that mixture expressed in grams of the second component in one hundred grams of the solution. Strictly speaking hydrobromic acid and water should not be considered as a mixture of two consolute liquids under atmospheric pressure because the solid hydrate HBr_2H_2O separates when the concentration of the hydrobromic acid passes a certain limit. It might be urged that all the haloid acids are gases in the pure state at ordinary temperatures under atmospheric pressure and this point is a sound one. On the other hand it has been deemed wise to include them in the table because they are very familiar examples of solutions with constant boiling points higher than those of either of the pure components. These solutions with constant boiling points have received a great deal of attention and the attempt was early made to treat them as compounds. Roscoe[1] showed that this could not be the case because the mixtures did not conform to the Theorem of Definite and Multiple Proportions and because the composition of the mixture with a constant boiling point was a function of the pressure under which the boiling took place. The mixture of hydrochloric acid and water which distills without change of temperature contains eighteen per cent. of hydrochloric acid when the external pressure is 180 $cm.$ of mercury and twenty-three and two-tenths per cent. when it equals 5 $cm.$ Hydrobromic and hydriodic acid behave in the same way, the solution containing more acid at lower temperatures and pressures. Nitric, hydrofluoric and formic acids show the opposite behavior, the amount of acid in the mixture with constant boiling point increasing with the external pressure. We are not able to predict the direction of this change. It is evident from the curves AA_1A, BB_1B that, with systems having a minimum or maximum boiling point, there can be two solutions with the same boiling point at any temperature between the minimum boiling point and the boiling point of the more volatile component in the one case and between the maximum boiling point and the boiling point of the more volatile component in the other case. A similar relation will hold in regard to the pressures at constant temperature. It has been pointed out by Gibbs[2] that the temperatures for which there can be

---

[1] Liebig's Annalen, **12**, 327 (1859); **116**, 203 (1860); Cf. Ostwald, Lehrbuch, I, 649.

[2] Trans. Conn. Acad. **3**, 156 (1876).

two solutions with the same pressure and the pressures for which there can be two solutions with the same temperature can be represented graphically. In the pressure-temperature diagram, Fig. 30, there is a qualitative reproduction of the behavior of propyl alcohol and water. The pressure-concentration curve BB_1B of Fig. 28 becomes a straight line ; B_2 is the vapor pressure of pure water at that temperature, B of pure propyl alcohol and B_1 of the solution having the same composition as the vapor. Suppose this to be done for several temperatures and curves drawn through all the B's, all the B_1's and all the B_2's. In the field below the B_2 line there can exist only vapor. In the field between the B_2 line and the B line occur all the simultaneous values of temperature and pressure for which only one pair of coexistent phases is possible. Between the B line and the B_1 line are all the simultaneous values of temperature and pressure for which there are two pairs of coexistent phases, while above the last line there can exist a liquid phase only.

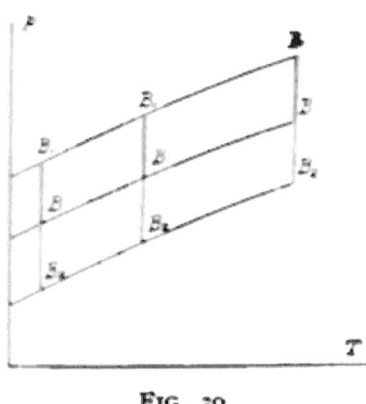

FIG. 30.

In the diagram, Fig. 29, the dotted line DD_1D is typical of all systems with intermediate boiling points which have yet been studied. As the position of the curve shows, the vapor always contains a greater percentage of the more volatile component than the liquid. Through fractional distillation, if carried on long enough, the two liquids will finally be separated completely, the more volatile being found in the distillate while the less volatile remains in the boiling flask. The Hempel column is merely an apparatus by which a large series of distillations is carried out in what seems to be a single operation. The vapor is partially condensed on the first set of beads, the uncondensed part becoming richer in the more volatile component, and this process is repeated through the whole length of the column. The condensed liquid flowing down through the tube washes off the beads so that the lower end of the column contains a liquid having a relatively large amount of the more volatile component and this takes yet more of the less volatile component out of the vapor. Other things being equal the column is more effective when in full action than when the distillation is first begun. The ease of separation depends on the difference between the percentage compositions of the vapor and the solution. If there is a large difference the two liquids can be separated easily; otherwise not. This can not be told from the boiling point curve alone, and it is not correct to state, as Ostwald[1] has done, that the ease of separation depends on the pitch of the curve. It is conceivable that there might be found a pair of consolute liquids which would give an almost straight boiling curve and yet have the composition of the two phases nearly identical at each temperature. On the other hand, it is to be noticed that when the boiling points of the two pure components lie very near together, the difference between the composition of the solution and vapor phases is apt to be small, and that such liquids can not readily be separated by fractional distillation.

It seems not impossible that the curve for the composition of the vapor phase might have the form DD_2D or DD_3D, though no example of this is known. Under these circumstances there would be one solution, represented by D_2 in the one case and by D_3 in the other, which would distill unchanged though the temperature is neither a maximum nor a minimum. Bauer[2] claims to have found that a solution containing equal reacting weights of propylene and ethylene bromide distills completely at 136°, a temperature between the boiling points of the two components. This statement has found its way into Beilstein's Handbuch;[3] but I am informed by Professor Orn-

---

[1] Lehrbuch, I, 648.
[2] Liebig's Annalen, Suppl. **1**, 250 (1861).
[3] Vol. I, 36.

dorff that it is not correct. A pair of liquids with the vapor composition curve DD,D would distill so that the mixture with constant boiling point would remain in the flask, and one or the other of the pure components would be formed in the distillate depending on the original composition of the solution. If the vapor curve were represented by DD,D the mixture with the composition D, would distill off, leaving one of the pure components in the flask. In both these cases we should have the very unusual phenomenon of the boiling point falling with continued distillation, something which is entirely outside of our experience. It is very much to be desired that some one should undertake a careful study of the vapor pressures, both total and partial, of mixtures of consolute liquids. At present the measurements of Linebarger are practically all that we have on the subject, and the accuracy of these is by no means unquestionable.[1] Lehfeldt[2] has made a few measurements on the composition of the vapors in equilibrium with different solutions at the same temperature; but he has not determined the vapor pressure. There is also one set of experiments by Winkelmann[3] on mixtures of propyl alcohol and water. No attention has been paid to the very curious vapor pressure curves found by Guthrie,[4] one of which certainly deserves careful examination, alcohol and amylene.

While it is impossible to say anything absolute about the behavior of two liquids when cooled below the temperature at which they are consolute, it is clear that there is a general connection between the form of the concentration-pressure curve and the nature of the phase which separates. The monovariant system, solid, solution and vapor, will give a curve with neither maximum nor minimum, if the system has a lower vapor pressure than that of one pure component and higher than that of the other. This is the usual case. The curve will have a maximum if the monovariant system has a higher vapor pressure than either pure component. If the monovariant system, two liquid phases and vapor, has a higher vapor pressure at the consolute temperature than either of the pure components, the

---

[1] Jour. Am. Chem. Soc. **17**, 615, 690 (1895).
[2] Phil. Mag. (5) **40**, 397 (1895).
[3] Wied. Ann. **39**, 1 (1890).
[4] Phil. Mag. (5) **18**, 515 (1884).

pair of consolute liquids will give a curve with a maximum pressure. This is the usual case, though there are no quantitative measurements illustrating it. If the vapor pressures of the monovariant system lie between those of the two components, the resulting divariant system will have all its pressures lying between those of the two pure components. As there are no experiments on this point it is not possible to state that such a curve has a change of curvature somewhere, though it is very probable. A minimum vapor pressure will occur if the solution on cooling forms the monovariant system, solid, solution and vapor, with a pressure less than that of either pure component. This can only happen when the solubility coefficient of each vapor in the other liquid is very large; but this is merely another way of saying the same thing and does not explain anything, because we are not in a position to make predictions in regard to the solubility coefficients.

Where there are two liquid phases in equilibrium, one of the components is solvent in the one phase, the other in the other. Each component has a definite and known solubility in the the other. At the consolute temperature the two phases have the same composition and become miscible in all proportions. At this temperature there

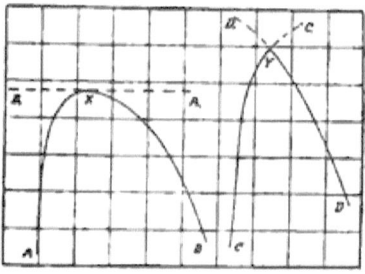

FIG. 31.

are two consolute liquids each with a definite and known solubility in the other, the two values being reciprocal. This raises the question of the form of the solubility curves at yet higher temperatures. The solubility of either liquid in the other may become infinite above the consolute temperature or it may not. The graphical representation of these two cases is given in the concentration-temperature diagram, Fig. 31. AXA_1 and BXB_1 are the solubility curves if the two solubilities become infinite at the consolute temperature; CYC_1 and DYD_1 are the solubility curves for a system in which no such change takes place at the consolute temperature, here represented by Y. At temperatures lower than Y, any mixture represented by a point in the field ACYD separates into two liquid phases whose concentrations are given by the points on CY and DY corresponding to the temperature of the experiment. Above the consolute temperature a mixture corresponding to any point in the field C_1YD_1 will not form two liquid layers because the mixture is unsaturated in regard to either component as solvent. In this field and in this field only is it a matter of complete indifference which component is taken as solvent. That such systems exist is shown by the fact that the curves CY and DY are exact representations of the behavior of sulfur and toluene.[1] There can be little doubt that the solubilities represented by these lines do not become infinite at the consolute temperatures. We see thus that it is not necessary for two liquids to have infinite solubilities one in the other in order to be miscible in all proportions; but that the same result is attained if the solubilities overlap.[2]

Many consolute liquids have definite but unknown solubilities, one in the other. The determination of these unknown solubilities is one of the problems yet to be solved. An examination of the pressure-concentration curves at constant temperature for mixtures of different alcohols and water, shows that there is probably a relation between the form of the curve and the solubility. It is well known that decreasing amount of carbon in homologous series means usually increasing solubility in water, and we find that, as we pass down the series, we have the isobutyl alcohol forming two liquid phases with water, giving the characteristic curve AA, Fig. 32. Propyl alcohol is miscible in all proportions with water, and the concentration-pressure curve BB has a distinct maximum; with decreasing amount of carbon we have ethyl alcohol and water, represented by CC; methyl alcohol and water, represented by DD. The curve CC is distinctly convex, as seen from above, while this is no longer the case with DD.[3] It should be kept in mind that the decrease in the

---

[1] Alexejew, Wied. Ann. **28**, 310 (1886).
[2] Cf. Horstmann, Graham-Otto's Lehrbuch, I, 32, (1885).
[3] Ostwald, Lehrbuch I, 647; Bancroft, Jour. Phys. Chem. **1**, No. 3 (1896).

reacting weights of the alcohols tends to modify the curves in the same way, and it is not impossible that the effect of increasing but unknown solubility may be very slight or even non-existent.

With salicylic acid and water it has been seen that a solubility curve may appear to be a continuation of a fusion curve and it is therefore not safe to assume that because the concentration-temperature diagram for a binary system has the general form of Fig. 27, the curve AB, for instance, is necessarily a fusion curve along its whole length. If it is a single curve, it must be a fusion curve because it starts from the melting point of one of the pure components

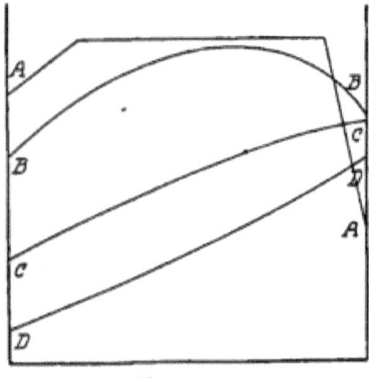

FIG. 32.

which no solubility curve can do. If, at any point, there is a sudden change of direction, there must be, at that point, the change from a fusion to a solubility curve with the other component as solvent if the precipitate remains unchanged. When it is remembered that the change of direction in passing from one curve to another may be infinitely slight and that no system has been studied with the object of determining whether such discontinuity occurs, it is not surprising that there is a deplorable lack of data in regard to this point. The experiments of Étard[1] are the only ones which are available. He finds a distinct break for silver nitrate and water at a temperature of about 45°, and for sodium nitrate and water at about 60°. Silver nitrate occurs in two modifications; but the change

[1] Comptes rendus, **108**, 176 (1889).

from one to the other takes place at 159.5° and does not seem to betray itself in the curve for the equilibrium between silver nitrate, solution and vapor. This is undoubtedly due to the large experimental error in the neighborhood of the fusion point. There is also a distinct break in the curve for triphenylmethane and carbon bisulfide[1]; but this may be due to the appearance of another modification of triphenylmethane since one is known.[2] It is rather strange that there should be no break when the carbon bisulfide is replaced by chloroform or hexane. Tilden and Shenstone[3] have found in barium acetate and water a mixture giving a curve which resembles that for the system, solid salicylic acid, solution and vapor. It it evident that there is a change with increasing concentration of barium acetate from a solubility to a fusion curve but it is impossible to tell from the experiments at what point this takes place. In all these cases the point at which the solution solidifies without change of temperature is the cryohydric point and not the eutectic point. It is probable that in few mixtures of salts and water is there an intersection of two fusion points; but the subject calls for a great deal more careful study than has yet been devoted to it. The formation of a true eutectic mixture by the meeting of two fusion curves is not confined to organic substances, though the bulk of the work has been done upon them as involving fewer experimental inconveniences.[4] It is probable that the mixtures of two salts which do not crystallize together are instances of true eutectic mixtures, provided the salts are consolute in the liquid state. In Table 25 are given the compositions of some binary salt solutions which solidify without change of temperature and the freezing points of these mixtures.[5] The compositions are expressed in reacting weights of the salt mentioned in one hundred reacting weights of the solution. The other salt is, in all cases, potassium nitrate which melts at 320°.[6]

---

[1] Comptes rendus, **115**, 950 (1892).
[2] Lehmann, Molekularphysik, I, 202.
[3] Phil. Trans. **175**, 23 (1884).
[4] Vignon, Comptes rendus, **113**, 133 (1881); Miolati, Zeit. phys. Chem. **9**, 649 (1892); Dahms, Wied. Ann. **54**, 486 (1895); Roloff, Zeit. phys. Chem. **17**, 325 (1895).
[5] Guthrie, Phil. Mag. (5) **17**, 643 (1884).
[6] Cf. Ostwald, Lehrbuch I, 1023.

TABLE XXV

	Temp.	Conc.
Potassium chromate	295°	2.0
Calcium nitrate	251	17.3
Strontium nitrate	258	14.3
Barium nitrate	278	14.0
Lead nitrate	207	21.2
Potassium sulfate	300	1.4
Sodium nitrate	215	36.8

If a solution containing two components, which can not crystallize together nor form two liquid phases, be cooled it will, at length, become saturated in respect to one component. If there is no supersaturation this will precipitate out and the temperature will fall until the point is reached where the other component separates in the solid form. From this moment the temperature will remain unchanged until the whole of the solution has disappeared. There will be only one point at which the temperature remains constant, though the rate of cooling will change when the first solid phase appears. If, however, the solution becomes supersaturated and cools a degree or two below the temperature at which the solid phase should appear, there will be a sudden rise of temperature when the solid phase actually separates out. If the mass of the liquid be fairly large and the thermometer not too sensitive, it will seem as if the temperature remained stationary for a short time. The temperature at which this takes place will change with the concentration of the solution whereas the temperature at which the solution finally disappears is independent of the original concentration. This phenomenon of the stationary and movable freezing point is really a very simple one, the movable point being the one that is determined in all cryoscopic measurements. In aqueous solutions, it has been familiar to everyone for years; but when it was noticed in melted alloys it aroused a great deal of interest and was the cause of many hypotheses.[1]

If the components form one or more compounds, the case will be different. If no one of the compounds has a true melting point,

---

[1] Cf. Ostwald, Lehrbuch I, 1018-1027. The first satisfactory treatment of the subject is due to Schultz, Pogg. Ann. **137**, 247 (1869).

there will be only one concentration at which the solution can solidify without change of temperature; while for every hydrate with a true melting point there will be two more such concentrations. In Fig. 33 are typical curves for the freezing points of binary systems in which the components can combine to form compounds but not solid solutions.

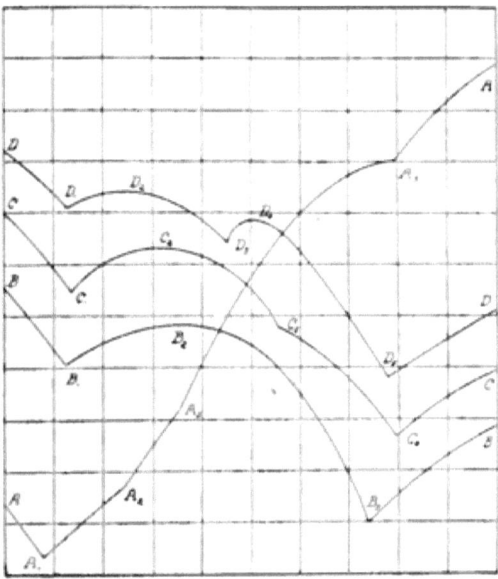

FIG. 33.

The curve $AA_1A_2A_3A_4A$ represents a system which forms three compounds, no one of which is stable at its melting point. This system is realized very nearly in the equilibrium between potassium hydroxide and water,[1] the three compounds being $KOH4H_2O$, $KOH2H_2O$ and $KOHH_2O$. As a matter of fact, the monohydrate reaches its melting point just before the curve for the anhydrous salt begins; but, ignoring that, there is only one concentration, represented by A_1, at which the solution will solidify without change of temperature, the solid phases being ice and $KOH4H_2O$. $BB_1B_2B_3B$

---

[1] Pickering, Jour. Chem. Soc. **63**, 890 (1893).

is a case exemplified by methylamine and water.[1] At B_1 the solution solidifies to ice and the trihydrate. B_2 is the melting point of the trihydrate, and a solution having that composition solidifies without change of temperature. At B_3 the solid phases are probably methylamine and the trihydrate, though this has not been determined experimentally. $CC_1C_2C_3C$ is the curve for a system forming two compounds, one of which has a true melting point. Trimethylamine and water illustrate this case, the two hydrates having the composition $(CH_3)_3N_11H_2O$ and $(CH_3)_3N_2H_2O$, though this latter compound has not been analyzed. The solution will solidify without change of temperature at C_1 to ice and $(CH_3)_3N_11H_2O$; at C_2 to the hydrate alone, and at C_3 to trimethylamine and the dihydrate. $DD_1D_2D_3D_4D_5D$ is a concentration-temperature diagram for a system forming two compounds, each stable at its fusion point. This case is realized in diethylamine and water where the solid hydrates are $(C_2H_5)_2NH_11H_2O$ and $H_2O_2(C_2H_5)_2NH$. As will be seen from the diagram, there are five concentrations at which the solution will solidify without change of temperature. It would be possible to make diagrams for still other hypothetical cases and find illustrations for them, as for instance, that there should be three compounds possible, of which the middle one only should not be stable at the fusion point. This would be represented by dimethylamine or ethylamine and water. One simple rule covers all cases. If the number of compounds stable at the melting point be "n," the number of solutions which will solidify without change of temperature is "$2n + 1$," if only stable states of equilibrium are considered. It follows that there can never be an even number of solutions which will do this, a rule which would be very serviceable in case two of the points were very close together, as they well might be.

The change of direction which occurs when a fusion curve passes into a solubility curve is not to be confounded with the "break" when a second modification of the solvent separates. In the latter case there is formed the nonvariant system, two solid modifications of the solvent, solution and vapor, and the temperature will remain constant so long as the four phases are present. In the solubility deter-

---

[1] Pickering, Jour. Chem. Soc. **63**, 141 (1893).

minations of Étard[1] there is a very distinct point of discontinuity in the curve for potassium nitrate and water at about 125°. Guthrie's measurements show the same phenomenon at about 130° though not so clearly.[2] Neither of them paid any attention to the nature of the crystals at this point; but there is little doubt that a second modification of potassium nitrate crystallizes out. Such a modification is known and its inversion temperature was found to be 129.5°.[3] In Fig. 34 is a graphical reproduction of Étard's results for silver and

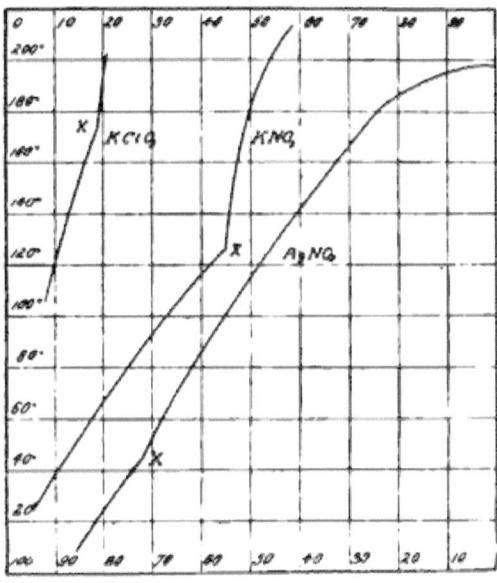

FIG. 34.

potassium nitrates and potassium chlorate. In the first case the break is due to the change of solvent, in the second case to a change in the nature of the solid phase and in the third case to one of the two. Potassium chlorate[4] occurs in two modifications but there are no satisfactory data as to the inversion temperature so that

---

[1] Comptes rendus, **108**, 176 (1889).
[2] Phil. Mag. (5) **18**, 114 (1884).
[3] Schwarz, Prize Dissertation, Göttingen (1892).
[4] Lehmann, Molekularphysik, I, 215.

it is impossible to say to what this change of direction is due. According to Étard there is a second break in the curve for potassium nitrate; but it is a little unsafe to draw any conclusions from it because Guthrie's measurements show no sign of it.

All the nonvariant systems containing two components which have been considered thus far have been made up either of two solid phases, solution and vapor or of one solid phase, two solutions and vapor. It is also possible to have a nonvariant system containing no liquid phase and composed of three solid phases and vapor. An example of this is to be found in the double salt of silver and mercuric iodides, $HgI,2AgI$. This salt changes color at 50° and it has been assumed that the change was analogous to the one taking place with mercuric iodide, the formation of another modification of the same salt.[1] Quite recently it has been shown by Bauer[2] that the double salt breaks up into its components forming the nonvariant system, mercuric iodide, silver iodide, double salt and vapor, an equilibrium which can exist at one temperature and one pressure only. In view of this, it is quite possible that the change in the double iodide of copper and mercury, which takes place at 88°, may also be a decomposition.

[1] Lehmann, Molekularphysik I, 169.
[2] Zeit. phys. Chem. **18**, 180 (1895).

CHAPTER IX

SOLID SOLUTIONS

There still remain to be considered the cases in which there can exist one or more solid phases with concentrations varying continuously within certain limits; in other words, the cases in which it is possible to have formation of solid solutions. No system of this class has been studied in detail, so that the discussion will have to be limited to a statement of the few facts already known and a reference to some of the equilibria yet to be realized. We have already seen that the freezing point of a solution is invariably lower than that of the pure solvent, if the latter crystallizes in the pure form. This is not necessarily true if a solid solution separates. The possibilities are best seen in the pressure-temperature diagram, Fig. 35. AO is the

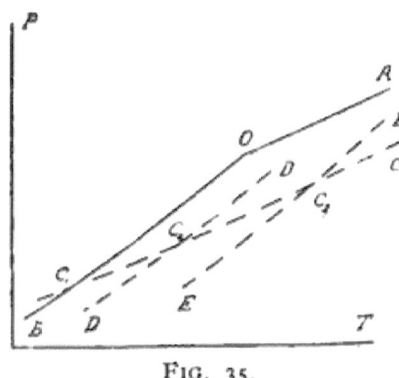

FIG. 35.

vapor pressure curve for the pure solvent as liquid, OB for the same as solid, CC_1 is the pressure curve for a solution. If the pure solvent crystallizes the freezing point of the solution will be at C_1, a temperature lower than O. If a solid solution separate, the partial pressure of the solvent in the solid solution will be less than the vapor pressure of the pure solvent at that temperature. The partial pressures of the solvent in the solid solution may be represented by the curve

DD if the solid solution is a dilute one, and by EE if it is a concentrated one. The liquid and solid solutions will be in equilibrium, so far as we know, when the partial pressure of the solvent in the vapor phase is the same for the two solutions.[1] The freezing point of a given solution is the temperature at which the pressure curve cuts the pressure curve for the particular solid solution with which it can be in equilibrium. In the two cases represented in the diagram these freezing points are at C_2 and C_3. Both these temperatures are higher than C_1, while C_3 is higher than O. The freezing point of a liquid solution is always higher if a solid solution separates than if the pure solvent crystallizes; if the solid solution is sufficiently concentrated, the freezing point is higher than that of the pure solvent. Both these cases have been realized experimentally. The first case is exemplified by thiophene and benzene,[2] m-cresol and phenol, iodine and benzene,[3] and a number of other mixtures.[4] It was to account for the abnormally small depressions of the freezing point that van 't Hoff[5] developed the conception of solid solutions. A rise of freezing point has been observed in many cases by Küster.[6] If the two components are miscible in all proportions in the solid phase, a nonvariant system is impossible since there can not be two solid phases by hypothesis, and it is improbable that two substances which are consolute in the solid phase will not be consolute in the liquid phase. Under these circumstances we shall expect to find, in the freezing points of such systems, all the types which were found for the boiling points of pairs of consolute liquids. There will be mixtures having a minimum freezing point lower than that of either of the pure components; there will be cases where the freezing point of some mixture is higher than that of either of the components, and, lastly, there will be instances where the freezing points of all mix-

---

[1] This leaves out of account the partial pressure of the solute which has not received a quantitative treatment. The most plausible assumption is that the partial pressures of the solute are equal for the two solutions when the system is in equilibrium.

[2] van Bijlert, Zeit. phys. Chem. **8**, 343 (1891).

[3] Beckmann and Stock, Ibid. **17**, 123 (1895).

[4] Ferratini and Garelli, Ibid. **13**, 1 (1894); Garelli, **18**, 51 (1895); Paternò, Ibid. **19**, 191 (1896).

[5] Ibid. **5**, 322 (1890). [6] Ibid. **8**, 577 (1891).

tures lie between those of the pure components. Only the last type has been studied as yet; but in this class there has already been found a limiting case which is entirely new. It is clear that if a solid solution has the same composition as the liquid from which it separates, the mixture will solidify without change of temperature. If, in a given system, the solid solution always has the same composition as the liquid phase from which it crystallizes, it follows that each mixture of these two components will behave like a homogeneous body with its own, definite, constant freezing point. An example of this has been found in the system composed of hexachlor-α-keto-γ-R-pentane and pentachlormonobrom-α-keto-γ-R-pentane.[1] The analogous case of two liquids, such that the composition of the vapor is always the same as that of the solution with which it is in equilibrium, has never been realized.

If the two components are not miscible in all proportions in the solid phase, there are several nonvariant systems possible. If each component is slightly soluble in the other solid component, there can exist the nonvariant system, two solid solutions, liquid solution and vapor. An instance of this seems to occur in the system composed of quinonedihydroparadicarboxylic ester and succinylosuccinic ester.[2] Thallium and potassium chlorates[3] probably offer another illustration of this equilibrium, though this has never been shown experimentally. If one component is soluble in the other, and the second insoluble in the first, there will be possible the nonvariant system, solid solution, one pure component as solid phase, solution and vapor. This has never been observed, but it seems probable that benzene and iodine come under this head. The two solid phases may also be a solid solution and a compound. If the two components form two sets of solid solutions and a compound or three sets of solid solutions, the possible nonvariant systems are increased, to say nothing of allotropic modifications or of the possibility of two liquid phases. It is a waste of time to attempt to classify these different equilibria until there are some experimental data upon the subject, and they are referred to here, chiefly, to call attention to our lack of knowledge.

[1] Küster, Zeit. phys. Chem. **5**, 601 (1890).
[2] Lehmann, Zeit. phys. Chem. **1**, 49 (1889).
[3] Roozeboom, Ibid. **8**, 531 (1891).

It has already been shown that when only the pure components or compounds can separate from the solutions, the number of mixtures for which the solution solidifies without change of temperature is $2n+1$ where n is the number of compounds with stable melting points. If a solid solution separates, this rule does not apply, and all that can be said is that if a solution solidifies without change of temperature, and the freezing point is not a maximum, there must have been formed two solid phases, one richer and one poorer in either component than the solution.

Passing to divariant systems we have to consider the absorption, adsorption or occlusion of gases by solids. These three terms probably refer to the same phenomenon in most cases and yet they are often used as though distinct. By many people absorption is used for the solution of gases and vapors in liquids, adsorption for the condensation of gases upon solids,[1] while occlusion often carries with it the idea of a hypothetical mechanical entanglement. These distinctions are rarely carried through consistently, and the solution of gases in metals is spoken of indifferently as absorption or occlusion. This general haziness in language is the sign of a corresponding lack of clearness in ideas. When a porous substance is brought into contact with a liquid, phenomena due to capillary action will certainly occur, and this may also take place when a gas is substituted for the liquid. On the other hand, capillary action will not account for all the phenomena observed, and it is rather doubtful whether it is an important factor in most cases. It would be well to keep the term "adsorption" for effects which may prove to be due primarily to surface tension, and to treat "absorption" as the general term applying to liquid and solid solvents, while "occlusion" would refer only to the formation of solid solutions. The conception of mechanical entanglement is to be given up as not describing the facts.

The solubility of gases in solids is much greater than is usually supposed, such different substances as wool, glass and metals being able to condense them.[2] Deville[3] first showed experimentally that hydrogen diffused through many metals; but his method was a

---

[1] Du Bois-Reymond, cf. Ostwald, Lehrbuch I, 1084.

[2] Cf. Ostwald, Lehrbuch I, 1084-1098.

[3] Cf. Lehmann, Molekularphysik II, 81.

rough one and yielded good results only at high temperatures. We owe to Helmholtz[1] a very delicate method based on the polarization of platinum by occluded hydrogen, and Thoma[2] has determined in this way which metals have the power of occluding hydrogen to any perceptible extent.

In applying the Phase Rule to these phenomena we are practically limited to the system, palladium **and hydrogen.** Since this is a divariant system at ordinary temperature and pressure, it must be possible for the system to have any pressure at **any temperature, and this** is the case experimentally. It also follows that for a given tem**perature and** pressure the concentration of the occluded **gas must** always have the same value. This is not true experimentally. Palladium foil and palladium black absorb different **amounts of hydrogen** under the same conditions. The explanation of **this anomaly is** not hard to find. Solids are rarely homogeneous. **There is a difference** between the surface and the interior, **and there are often stresses** throughout the mass due to the methods **of preparation.** Such a solid is not really **in** equilibrium; **but the passive resistances to** change are so great, or the reaction velocity **is so low, that the final** equilibrium may never be reached. **If two pieces of metal are not** exactly alike there is no reason that the **absorption coefficients should** be identical. This may seem like **a far-fetched explanation; but** any one, who has compared two pieces of **metal electrically, knows** what differences may and do exist between **different parts of the same** rod. Under the circumstances the values for the **occlusion of hydrogen** by palladium are not absolute **because they depend upon the** sample of metal used, or rather **upon the treatment it has received.** By always using metal prepared under **the same conditions, it is possible** to obtain comparable values.

From a study of the relation between **the pressure of the vapor** phase and the concentration in the solid phase, **van 't Hoff[3] decided** that the compound Pd_2H was first **formed, and that the hydrogen** occluded in excess of this formed **a solid solution of hydrogen in pal-**

[1] Ges. Abh. I, 835.
[2] Zeit. phys. Chem. **3**, 69 (1889).
[3] Ibid. **5**, 128 (1890).

ladium hydride. The careful measurements of Roozeboom and Hoitsema[1] have failed to confirm this view. They find that the pressure-concentration diagram is composed of three parts. With increasing concentration of hydrogen in the solid phase the pressure increases up to a certain value. Beyond this point the pressure remains nearly constant while the concentration in the solid phases increases. When this latter has reached a certain value, the pressure and concentration, vary again simultaneously. This form of curve was entirely unexpected, and it is not certain how it is to be explained. Hoitsema is in doubt whether this points to a condensation of hydrogen to the liquid form, or to the formation of two solid solutions. Neither of these explanations is very satisfactory and the matter must be left open for the present, the more especially since the latest work on the occlusion of hydrogen by platinum[2] has not yet led to any definite conclusion in regard to the nature of the solid phase or phases present.

In the case of the occlusion of gases by carbon there is little doubt that we have the formation of a solid solution. The experiments of Chappuis[3] bring out the striking analogy between the absorption of gases by liquids and by solids. The different kinds of charcoal showed very different absorption powers; but this is not surprising when one considers the variations in chemical properties under the same circumstances.[4]

[1] Zeit. phys. Chem. **17**, 1 (1895).
[2] Mond, Ramsay and Shields, Ibid. **19**, 25 (1896).
[3] Wied. Ann. **12**, 161 (1881).
[4] Cf. Meslans, États allotropiques, 112.

CHAPTER X

REVIEW

Before passing to systems made up of three components, it will be well to give a brief summary of the different kinds of nonvariant systems studied, with an illustration of each, so far as this is possible. In the nonvariant systems made up of two solid phases, solution and vapor, the two solid phases are the two components as solvents in the case of naphthalene and phenanthrene; the two components, one as solvent, the other as solute, in the case of ice and potassium chloride, ice and naphthalene; solute and compound in the case of sodium sulfate and hydrated sodium sulfate; two compounds in the case of the ferric chlorides with twelve and seven of water or iodine monochloride and trichloride; two modifications of the solvent in the system, potassium nitrate and water; two modifications of the solute in the system, sulfur and toluene; two solid solutions in the case of quinonedihydroparadicarboxylic ester and succinylosuccinic ester, or possibly potassium and thallium chlorates. The cases where the two solid phases are a compound and solid solution or solvent and solid solution, have not been realized with certainty, while the case of solute and solid solution probably occurs with iodine and benzene, though this lacks experimental confirmation.

There have been two kinds of nonvariant systems studied, in which there are two solutions, solid and vapor. In one the solid phase is one of the components, a case realized by water and naphthalene; in the other the solid phase is a compound, exemplified in the equilibrium between sulfur dioxide and water. In the nonvariant system, three solid phases and vapor, there can be the two components and a compound as solid phases. This has been discovered in the system composed of silver iodide and mercuric iodide. There are no cases known where a solid solution is one of the three solid phases, while a nonvariant system composed of three liquid phases and vapor is probably impossible.

Since metals are not soluble in ordinary solvents, they are often looked upon as forming a class by themselves, and it is tacitly assumed in many cases that the behavior of alloys is not described by the theorems applicable to ordinary chemical phenomena. This is a mistake. All the conclusions in regard to binary systems which have been reached in the previous discussion might have been illustrated by taking suitable pairs of metals. Addition of one metal to another lowers the freezing point of the second if the pure solvent separates.[1] If two metals form neither compounds nor solid solutions and are consolute in the liquid form, the nonvariant system, two solids, solution and vapor, will be possible at one temperature and pressure only, and that temperature will be lower than the freezing point of either of the pure components. If a mixture of two such metals be heated until completely liquefied and the molten mass allowed to cool slowly, a sudden change in the rate of cooling will be noted at a temperature which varies with the original composition of the solution ; at a lower temperature the temperature will remain constant until all the metal has solidified. The change of rate occurs when one of the metals crystallizes from the solution and is therefore a function of the concentration. The constant temperature comes at the eutectic temperature.[2] Examples of this are lead and silver, lead and bismuth, tin and bismuth, zinc and tin. The fact that one of the metals separates in the pure state until a certain concentration is reached, is made use of technically in the Pattinson process to enrich silver ore.[3] An ore rich in lead and poor in silver is melted and allowed to cool to the eutectic temperature when the liquid is poured off. During the cooling, pure lead crystallizes and the concentration of silver increases to the value corresponding to the eutectic alloy. Lead and tin also form an eutectic alloy ; but Kopp[4] states that at high temperatures these metals cease to be consolute and form two liquid layers. Unless there is an error in the determination, this is analogous to the case of diethylamine and

---

[1] Cf. Tammann, Zeit. phys. Chem. **3**, 441 (1889) ; Heycock and Neville, Jour. Chem. Soc. **55**, 666 (1889) ; **57**, 376 (1890) ; **61**, 888 (1892).

[2] Cf. Ostwald, Lehrbuch I, 1025.

[3] Cf. Guthrie, Phil. Mag. (5) **17**, 466 (1884).

[4] Liebig's Annalen, **40**, 184 (1841).

water, where two liquid layers are formed on heating. Instances corresponding to naphthalene and water, where there may be equilibrium between two liquid phases, solid and vapor, are to be found with lead and zinc, zinc and bismuth, bismuth and silver.[1] Very recently it has been shown that zinc and bismuth become consolute at about 825°; zinc and lead above 900°.[2] Since one characteristic feature of a nonvariant system is that the temperature remains constant until one of the phases has disappeared, it is clear that mixtures of two metals which form two liquid layers will show two points in cooling at which the thermometer reading will remain constant for a time. The temperatures of these points are independent of the initial concentrations, provided two liquid layers are formed. If either of the components happens to be present in very small quantities, the nonvariant system, solid, two solutions and vapor, may not be formed. With solutions of metals it is often impossible, owing to experimental difficulties, to tell by inspection whether two liquid phases are or are not formed; but a study of the rate of cooling will answer this question at once.[3]

Metals form definite compounds, gold and aluminum, silver and aluminum being examples of this. The complete curve for silver and aluminum has recently been determined by Gautier.[4] In this case the compound, Al_4Ag, is stable at its melting point. This is also true of the compound formed of gold and aluminum, which has a melting point higher than that of either component. There are probably many instances of definite compounds which cease to be stable before the melting point is reached; but the work on alloys has been done, for the most part, in such an unsatisfactory way that it is very difficult to tell what the facts are. To take a single illustration, the existence of a compound NaK is assumed, because a solution containing these two metals in the proportions corresponding to the formula solidifies without change of temperature.[5] If the temperature at which this takes place is really the melting point of

[1] Lehmann, Molekularphysik I, 572.
[2] Spring, Zeit. anorg. Chem. **13**, 29 (1896).
[3] Schultz, Pogg. Ann. **137**, 247 (1896).
[4] Comptes rendus, **123**, 109 (1896).
[5] Hagen, Wied. Ann. **19**, 436 (1883).

this compound, there must be two other solutions which will also solidify without change of temperature, yet this point has not been investigated. It is not at all impossible that this compound does not exist, and that people have been misled by the composition of the eutectic alloy happening to coincide very closely with the composition of a definite compound.

Metals also form solid solutions, and Tammann's experiments with mercury and potassium, mercury and sodium point to the existence of a nonvariant system with mercury and amalgam as the two solid phases in both cases.[1] Heycock and Neville[2] observed the phenomenon of a rise of freezing point with silver and cadmium, antimony and bismuth, antimony and tin, while Tammann noticed the same behavior for gold, tin and cadmium in mercury, and Gautier[3] seems to have found a case with antimony and aluminum, where the addition of the more fusible to the less fusible metal raises the freezing point. The pair of metals, antimony and tin, has been studied in detail by van Bijlert[4] and by Küster.[5] It was found that the two metals are miscible in all proportions in the solid phase, and that a nonvariant system is impossible. Küster complicates matters unnecessarily in this paper and elsewhere by making a distinction between isomorphous mixtures and solid solutions. With copper and nickel the general form of the freezing point curve makes it probable that the solid phases at one inversion point are two sets of solid solutions.[6]

Some other determinations by Gautier are more difficult to interpret because one does not know the probable error of his measurements. With nickel and tin, the curve passes through a maximum and the composition of the solution at this point does not correspond very closely to that of any definite compound. Gautier himself thinks that the discrepancy is due to experimental error; but in view of the fact that in the silver and aluminum series, the maximum comes

[1] Zeit. phys. Chem. **3**, 441 (1889).
[2] Jour. Chem. Soc. **61**, 911 (1892).
[3] Comptes rendus, **123**, 109 (1896).
[4] Zeit. phys. Chem. **8**, 343 (1891).
[5] Ibid. **12**, 508 (1893).
[6] Gautier, Comptes rendus, **123**, 172 (1896).

very sharply at the right place, it is more probable that with nickel and tin there is formed a solid solution which can coexist at one point with a liquid solution of the same composition. When nickel is added to tin, there is at first a depression of the freezing point ; but this extends over a very narrow range of concentrations. Whether the solid phase is pure tin along this portion of the curve is open to doubt. The most natural assumption would be that the order of crystallization was tin, solid solution, and lastly nickel ; but there is no parallel to this in any carefully studied case, so that it is not impossible that solid solution separates from the beginning, and that the first minimum freezing point does not denote the existence of a nonvariant system. With antimony and aluminum there are two maxima, each higher than the melting point of either component. It is probable from the experiments that neither is the melting point of a compound, though it is unsafe to draw any conclusions because it is stated that some of the crystals, after standing, do not melt at the highest temperature reached by any portion of the freezing point curve. All the measurements with antimony as one component need revision, because Gautier seems to have worked entirely with what is usually called amorphous antimony.[1]

---

[1] Cf. Meslans, États allotropiques des corps simples.

THREE COMPONENTS

CHAPTER XI

GENERAL THEORY

With three components, five phases are necessary to constitute a nonvariant system; four for a monovariant, and three for a divariant system. It will simplify matters to consider first, the cases where there can be only one liquid phase and the solid phases vary in composition discontinuously, taking up next the instances where a solid solution can be formed, and finishing with systems in which two liquid phases can be in equilibrium. The change of the pressure with the temperature can be represented equally well whether the system under consideration be composed of two, three, or any number of components; but a concentration-temperature diagram presents great difficulties when the number of components equals three. The problem has been solved in quite a number of ways. Most of the methods give a solid figure, the temperature being taken as the vertical axis; but it is possible to tell a great deal from the projections of the curves for the monovariant systems upon a plane, even though the temperature can no longer be read directly.

Schreinemakers' takes for the X and Y axes the amounts of two of the components in a constant quantity of the third. This is open to the objection that there is no place in the diagram for an anhydrous double salt, nor for solutions containing very little of the third component. Meyerhoffer[1] has invented a diagram which has the merit of allowing one to take the temperature as one of the co-ordinates. In a system composed of two salts and water, he measures the ratio of one salt to the other along one axis and the temperature along the other. This is serviceable in certain cases; but is very limited in application, since it neglects the relative quantities of both

---
[1] Zeit. phys. Chem. **9**, 67 (1892).
[2] Ibid. **5**, 97 (1890).

salts in respect to the third component. The method proposed by van Rijn van Alkemade[1] seems to have no advantage over the diagram of Schreinemakers. Gibbs[2] has suggested the use of a triangular diagram, the sum of the components being kept constant. If we take an equilateral triangle of unit height, the corners of the triangle will represent the pure components, and any point within the triangle will represent some definite mixture of the three substances. The amount of each component is given by the length of the perpendicular from the point to the side opposite the vertex corresponding to that component. This diagram has been used by Thurston[3] in some works on alloys, and was also suggested independently by Stokes.[4] Roozeboom[5] has used a modification of this diagram. He takes the isosceles right-angle triangle, the equal sides being of unit length. The advantage of this arrangement is that one can use the ordinary co-ordinate paper; but it is open to the objection that there is a different scale along the hypotenuse from that along the sides, so that one of the components seems to occupy an exceptional position. While this is not serious in the case of two salts and water where the water is solvent and the salts solutes, it is a disadvantage in the ternary systems in which no such distinction exists and becomes impossible when the system of three components is considered as a subdivision of one containing four.[6] Roozeboom[7] has proposed another form of triangular diagram which is distinctly superior to any of those already considered. It consists of an equilateral triangle with lines ruled parallel to each side instead of perpendicular. The length of one side is taken equal to unity or one hundred, and the same scale is used for the binary systems in the side of the triangle as for the ternary systems in the interior. Since co-ordinate paper can now be obtained, ruled in three directions,[8] this method of

---

[1] Zeit. phys. Chem. **11**, 306 (1893).
[2] Trans. Conn. Acad. **3**, 176 (1876).
[3] Proc. Am. Ass. **26**, 114 (1877).
[4] Proc. Roy. Soc. **49**, 174 (1891).
[5] Zeit. phys. Chem. **12**, 369 (1893).
[6] Ibid. **15**, 147 (1894).
[7] Ibid. **15**, 145 (1894).
[8] Cf. Bancroft, Jour. Phys. Chem. **1**, No. 7 (1897).

plotting results will be used except in special cases where some point is to be brought out not involving the variation of all three components.

Since no single ternary system has been studied in detail, it is impossible to select typical cases illustrating different points, as was done when there were only two components. The special cases, which have been worked out experimentally, have been selected chiefly for their complexity, and are not well adapted to bringing out the more general relations. All that can be done is to point out the general form of the boundary curves in the more simple cases, and to go over what experimental data there are. Excluding solid solutions we can have three different classes of solid phases, a pure component, a binary compound and a ternary compound. In Fig. 36 are the boundary curves for various possible combinations, starting with the simplest case. All solutions contain one hundred reacting weights of $A + B + C$. When the three components form no com-

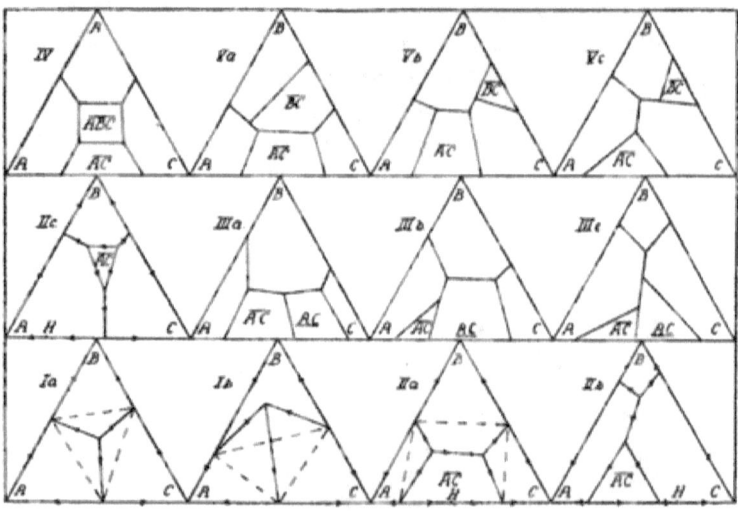

FIG. 36.

pounds, the boundary curves consist of three lines meeting in a point. The usual form is represented by Ia. The component A exists as solid phase in the field marked A, the components B and C in the fields marked with those letters. The line separating the fields A and C gives the compositions of the solutions which can be in equilibrium with A and C as solid phases. Along the two other lines the solid phases are A and B, B and C, respectively. At the intersection of the three lines there exists the nonvariant system, three solid phases, solution and vapor. It is possible to say something in regard to the temperature changes. The corners of the triangle represent liquid phases, each composed of one of the three pure components in equilibrium with that component as solid phase. The temperature of the lower left hand corner of the triangle is that of the melting point of A; at the upper corner is the melting point of B, and at the lower right hand corner the melting point of C. The point on the side AC from which the line starts toward the center of the triangle represents the composition of the eutectic alloy of A and C, and the temperature at this point is that at which this alloy freezes. As this temperature is always lower in such cases than the freezing point of either of the pure components, the temperature must fall as we pass along the side of the triangle from this point to the apex A or the apex C. This is shown in the diagram by the arrow heads which point in the direction of rising temperature. The same reasoning applies to the other sides of the triangle and the results are expressed in the same way.[1]

A theorem by van Rijn van Alkemade[2] will serve as a very effective guide in regard to temperature changes in the interior of the triangle. If the two points in the triangle which correspond to the compositions of two solid phases be connected by a line, the temperature, at which these same two phases can be in equilibrium with solution and vapor, rises as the boundary curve approaches this line, becoming a maximum at the intersection though the boundary curve often ceases to be stable before this point is reached. When the two solid phases are two of the components, the line connecting the melting points is one of the sides of the triangle. It is therefore clear that the temperature must always rise in passing along a boundary curve to the side of the triangle, if the theorem of van Alkemade be right. So far only one exception is known, and the measurements

---

[1] Cf. Roozeboom, Zeit. phys. Chem. **12**, 371 (1893).
[2] Ibid. **11**, 289 (1893).

in regard to this point were not all made by the same man. Until the number of contradictions is somewhat increased or until it has been shown under what circumstances the theorem does not apply, it may be accepted provisionally as accurate. It should be mentioned that Schreinemakers[1] has reached the same conclusion in regard to temperature changes along boundary curves terminating at the sides of the triangle.

In the particular diagram under discussion, Ia, the point representing the composition of the solution in equilibrium with three solid phases and vapor lies within the triangle formed by the dotted lines connecting the three binary eutectic alloys. While it is quite possible that this is necessarily the case, there is no conclusive proof of it and Ib may represent an actual system. Here the point O lies outside of the triangle formed by the dotted lines and, in this particular case, the nonvariant system formed from A, B and C would probably exist at a higher temperature than either of the nonvariant systems formed from A and B alone or B and C alone. The directions of the temperature changes are shown, as in the preceding diagram, by the arrow heads. Since no system behaving like this has yet been found and since it has as yet no theoretical justification, it will be unnecessary to consider the diagram derived from it when the three components can form binary or ternary compounds. In the following discussion only those cases will be considered in which all the boundary curves lie within the figure formed by connecting the points at which these curves meet the sides of the triangle.

Starting from Ia and making the assumption that one compound is possible we shall get the diagram for the new system by drawing a line from any point on any side to any point on any boundary curve or by drawing a line from one boundary curve to another. To take a concrete case, let us suppose that A and C form a compound having the formula A_xC, which we will represent in the diagram by AC. There are then three possible cases, shown in IIa, IIb and IIc. The binary compound $\overline{AC}$ and the component B can occur simultaneously as solid phases (see IIa); $\overline{AC}$ and B cannot occur simultaneously as solid phases (see IIb); $\overline{AC}$ and B can occur

---

[1] Zeit. phys. Chem. **12**, 73 (1893).

simultaneously as solid phases but AC cannot exist as a solid phase in the binary system composed of A and C (see IIc). The first two cases are the more common since IIc can only occur when the components A and C can form the nonvariant system, two pure components, a compound and vapor. As only one instance of this is known, in the double salt $AgIHgI_2$, the diagram IIc can only be realized at present by a study of the system, silver iodide, mercuric iodide and a third substance which melts below 50° and in which the other two are soluble.

In IIa and IIb the change of temperature along the side AC will be different from that in Ia if the composition of the binary compound AC is represented by a point lying between the two intersections of the boundary curves with that side of the triangle. This has been assumed to be the case in IIa, the composition of the double salt being given by the point H. As the compound can then exist in equilibrium with a solution of the same composition as itself the temperature at H is that of the melting point of the salt and is higher than that at the intersection of either boundary curve with the side of the triangle. In IIb the diagram has been so drawn that the binary compound has no true melting point, since H lies outside the field for AC. In this case there will be only one temperature minimum along the side of the triangle, at the point where the boundary curve for A and AC meets the side of the triangle. In IIa the boundary curve separating the field for B from that for AC will pass through a maximum temperature because it is cut by the line BH connecting the melting points of B and AC. In IIb the temperature of the nonvariant system with the double salt as one component will be higher than that of the nonvariant system with the component B as solid phase because A and C are solid phases along the boundary curve connecting these two points and the temperature therefore rises as one approaches the side AC of the triangle.

In the diagrams IIa and IIb there is no difficulty in telling what two components unite to form a binary compound; but in IIc there is no way of knowing whether the compound is made up of A and B, B and C, or A and C. It is probable that in any actual case the diagram would not be symmetrical, and one might be able to draw some conclusions from the irregularity; but this is purely hypothetical. Roozeboom[1] has pointed out that the theorem of van Alkemade requires that the temperature shall rise along two sides of the triangular field for the compound to the common vertex, and that this vertex points towards the side occupied by the two components which form the compound. In this way it is possible to determine the constituents of the compound without an analysis of the solid phase. The results given by inspection will be qualitative only. If the line connecting B and H in IIc cuts the boundary curve along which B and AC are solid phases, that point will be a maximum temperature for the boundary curve. If this is not the case, the temperature will be highest at the end of the boundary curve near BH. It is to be noticed that in IIa and IIb there is one curve which does not reach the side of the triangle, and that in IIc there are three such. Roozeboom has proposed calling these "middle" curves and the others "side" curves.[2]

Next in order of simplicity is the ternary system in which A and C form two compounds, AC and AC, while there are no compounds containing B. Since the system illustrated by IIc is a very unusual one, it is hardly worth while to consider the probability of A and C forming another compound. Excluding this, the diagram for all possible systems satisfying the requirements will be obtained by drawing a line from any point on AC to any one of the boundary curves in IIa or IIb. The three types of curves thus obtained are shown in IIIa, IIIb and IIIc. It should be mentioned that there is always a change of direction when one line meets another. This is not shown in all the figures because the direction of the change is not always known. In IIIa either compound can exist simultaneously with B as solid phase; in IIIb only one can, and in IIIc neither can. The directions of the temperature changes along the boundary curves separating the field for B from the fields for the compounds can only be told when the compositions of the compounds are known. As it is improbable that three components can form a ternary compound when no two can form a binary compound, we can pass at once to

[1] Zeit. phys. Chem. **12**, 384 (1893).
[2] Ibid. **12**, 363 (1893).

the case where there is formed one binary compound $A\bar{C}$ and one ternary compound $AB\bar{C}$, illustrated in IV. Since all substances are somewhat soluble in all liquids it follows that a ternary compound $AB\bar{C}$ can exist only in equilibrium with solutions containing some of all three components, and that its field will be bounded entirely by middle curves.[1] If the compound $AB\bar{C}$ has a true melting point each curve bounding the field for ABC may have a maximum temperature, and at least one must have. If it has not a true melting point, the four middle curves can not all have maxima. and it may happen that none have. There are three types of curves when there are two binary compounds not having the same constituents. In Va there is one nonvariant system possible with two of the components as solid phases; in Vb there are three such, and in Vc there are two such and one in which the three components form the three solid phases. The systems possible when there are three binary compounds AC, BC, and AB will be found in the same way by drawing lines from the side AB of the triangles under V to the other lines in the triangles; but the number of hypothetical cases becomes so large that it does not seem worth while to take them up individually, while our knowledge of the subject is not yet sufficient to permit of an exhaustive treatment of the possible forms when there can exist two binary compounds $A\bar{C}$ and $B\bar{C}$ together with one or more ternary compounds. At present we have no clue to enable us to distinguish between cases which can actually exist and purely artificial forms which are the geometrical consequences of a certain distribution of lines in a plane. Instead of further speculation in this direction, it will be more profitable to consider the experimental data which have already been obtained.

Guthrie[2] has studied the behavior of mixtures of potassium, sodium and lead nitrates so that it is possible to construct a diagram which shall represent the facts with some approach to accuracy. This is done in Fig. 37. The quantities are expressed in reacting weights, the sum of the three components being always equal to one hundred. A is the corner for potassium nitrate with a melting point,

---

[1] Roozeboom, Zeit. phys. Chem. **12**, 387 (1893).
[2] Phil. Mag. (5) **17**, 472, (1884).

according to Guthrie, of 320° though the more accurate measurements of Carnelley and others[1] point to about 340° as the true temperature at which potassium nitrate fuses. B is the corner for lead nitrate but the temperature of the point is unknown since this substance decomposes before melting. C is the corner for sodium nitrate and represents a temperature of 305° on Guthrie's thermometer. At D potassium and sodium nitrates are present in the proportions corresponding to the eutectic alloy and the temperature of

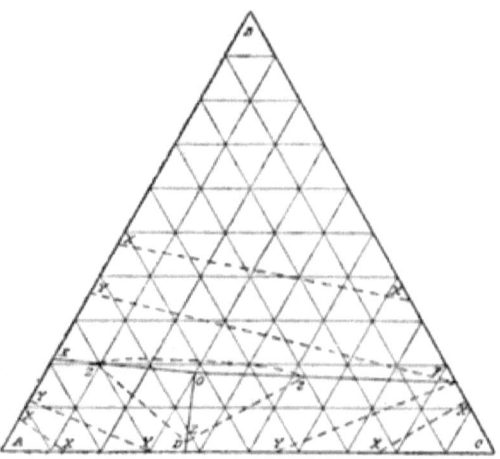

FIG. 37.

this point is 215°. E represents the eutectic alloy of potassium and lead nitrates with a temperature of 207°, while F is the corresponding point for lead and sodium nitrates, the temperature being 268°. Along DO potassium and sodium nitrates are the solid phases; along EO potassium and lead nitrates and along FO lead and sodium nitrates. At O there exists the nonvariant system, potassium, sodium and lead nitrates, solution and vapor. The temperature at which this system alone can exist is 186°. In the field ADOE there exists the divariant system, solid potassium nitrate, solution and vapor. In the solution potassium nitrate is solvent and the other two components are solutes. In the field CDOF there exists the divariant system, solid sodium nitrate, solution and vapor,—sodium

---

[1] Landolt and Börnstein's Tabellen, 148.

Three Components

nitrate being solvent in the solution. In the field **BEOF** there exists, theoretically, the divariant system, solid lead nitrate, solution and vapor. Practically this is interfered with by the partial decomposition of the lead nitrate. This is a secondary phenomenon and makes the selection of lead nitrate as one of the three components an unfortunate one.

It will be interesting to consider the general form of the isothermal curves for a system like this one, made up of three components which form no compound. It is explicitly assumed, for purpose of discussion, that lead nitrate **is stable**. The dotted lines X X represent the general form of the isotherm **for a temperature higher than that of** any of the binary eutectics **and lower than the melting point of the** most fusible component, **in other words for a temperature of** about 290°. The lines are not **really straight lines any more than** DO, EO and **FO are; but they are drawn approximately straight** because the only **determination we have in the interior of the triangle is that of the point O.** It will be noticed that this particular isotherm consists of **three parts**[1] **because the temperature rises in both directions along the sides of the triangle from the binary eutectics. At** the temperature **of the eutectic alloy of lead and sodium** nitrates, 268°, the **two branches of the curve come together on** the BC side of the triangle **and the isotherm for 268° has the** general form shown by the **dotted lines YY and YFY. At a lower** temperature two branches **of the isotherm would meet at some point** on the curve OF without ever **reaching the side BC. At a temperature lower than the freezing points of any of the binary eutectics, at** 195° for instance, the general form of **the isotherm will be given by** the dotted lines **ZZZ.** The isotherm **has become a closed curve which** will diminish in area as **the temperature falls until, at 186°, it will** consist only of the point **O. At still lower temperatures there will** be no isotherm because there will **be no solution and the solid phases do not react** each with **the other. The** isotherms approach the point O with falling temperature and **recede from** it with rising temperature. When the temperature **rises above the** melting point of the

---

[1] Roozeboom treats each part as a whole, speaking of the isotherm for a given solid phase; but, this is not to be commended.

most fusible compound, the isotherm consists of two branches only because a component can not exist as solid phase at a temperature above its own melting point. At a temperature below the melting point of the least fusible component and above those of the other two, the isotherm has only one branch while it is reduced to a point at the melting point of the least fusible component. Above this last temperature there can be no isotherm because the three components are liquids and all consolute, by hypothesis, so that a divariant system is impossible.

In considering the change of the isotherms with the temperature for a ternary system which permits of no compounds and no second solution phase, it is necessary to distinguish two cases. The melting points of the pure components may each be higher than the melting point of any of the binary eutectics. This is true with potassium, sodium and lead nitrates, and is the rule when the three components are similar in nature. One of the components may melt at a lower temperature than the binary eutectic formed from the other two components. This would occur with ice and two anhydrous salts, and is the rule whenever the freezing point of one component lies far below the freezing point of either of the other two. In the first case, the isotherm for a monovariant system will pass through the following forms as the temperature falls: a point, a branch and a point, two branches, two branches and a point, three branches, an isolated and two connecting branches, three connecting branches, a closed curve, a point. In the second case, the order will be: a point, a branch, a branch and a point, two branches, two connecting branches, two connecting branches and a point, two connecting branches and a branch, three connecting branches, a closed curve, a point. The limiting points for the isotherms are the highest temperature at which any solid phase can exist in equilibrium with solution and vapor and the lowest temperature at which this is possible.

In the case of the melted nitrates, the field in which any one of the components is solvent is identical with the field in which that component can occur as solid phase; but this is not necessarily true. As a rule this will be so only when the melting points are not too far apart. It would be an interesting point to consider whether the change came when the melting point of one of the components fell below that of the eutectic **formed from the other two**; but there are as yet no data on which **to base an opinion. The most** striking **case** of a substance **existing as solid phase without being the solvent in** the solution is **to be found in the equilibrium between two salts and water. The two salts must have the same acid or the same basic radicle, otherwise the number of components is no longer three. In Fig. 38 is the graphical representation of what few data are accessible for the system potassium chloride, potassium nitrate and water.**

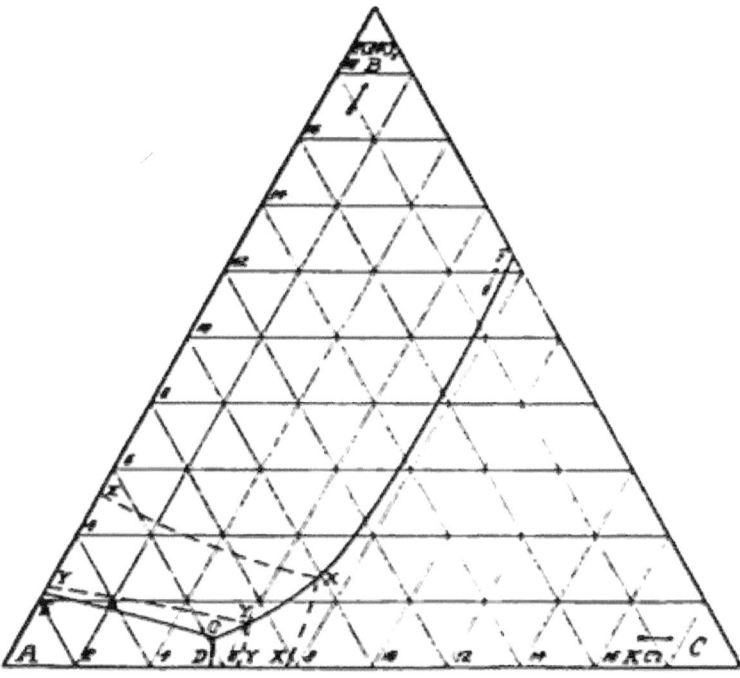

FIG. 38.

Only a small portion **of the diagram is shown, as the whole would** require a triangle five times as long on **the side. The concentrations are expressed in reacting weights, the sum of the three components being always equal to one hundred reacting weights. At D is the cryohydric point for ice and potassium chloride**[1]**, the temperature**

[1] Guthrie, Phil. Mag. (5) I, 49. (1876).

being $-10.7°$. At E, $-3°$, is the cryohydric point for ice and potassium nitrate. Along EO ice and potassium nitrate are solid phases; along DO ice and potassium chloride and along FO potassium nitrate and sodium nitrate. The concentration and temperature corresponding to the point F have not been determined experimentally Étard[1] has stated that the line OF terminates at the melting point of the more fusible salt. In other words, in this particular case, the amount of water in the solution in equilibrium with solid potassium chloride and nitrate will become zero at the melting point of potassium nitrate. This is entirely wrong. The curve OF terminates at the temperature of the eutectic alloy formed from the two salts, a temperature which is necessarily lower than the melting point of potassium nitrate. Curiously enough,'Étard has an inkling of the truth in one case[2], but it is not sufficient to make him modify his erroneous hypothesis. At O there exists the nonvariant system, ice, potassium chloride, potassium nitrate, solution and vapor, five phases. The temperature is $-11.4°$. In the field ADOE ice is solid phase; in BEOF potassium nitrate and in CDOF potassium chloride. Water, however, is solvent in the whole of the first field and in parts of the two others. Where the change takes place is not known; but there would be a line starting from the point where EB ceases to be a solubility curve and becomes a fusion curve and this line would cut OF at some point. On one side of this hypothetical line water would be solvent; on the other potassium nitrate. Similarly a line could be drawn from a point on DC to OF, such that on one side of it water would be solvent, on the other potassium chloride. The determination of these lines is one of the interesting problems in theoretical chemistry.

The isotherm for $20°$ has the general form shown by the dotted lines XXX.[3] It consists of two connected branches. A third branch is impossible because the temperature is above the melting point of ice. The isotherm for $0°$ consists of two connected branches and a point, illustrated by the dotted lines YYY and the point A. At a

---

[1] Comptes rendus, **109**, 740 (1889), Ann. chim. phys. (7) **3**, 275 (1894).
[2] Ibid. (7) **3**, 284, (1894).
[3] Nicol, Phil. Mag. (5) **31**, 369 (1891).

little lower temperature the point will become a line across the corner and at temperatures below $-10.7°$ the isotherm will have the form of a closed curve. In all cases where the isotherm crosses a boundary curve, there is a change of direction because there is a change in the nature of the solid phase. There is also a change of direction when the isotherm cuts the line separating the field for one component as solvent from the field in which another component is solvent. It matters not whether one is dealing with a system of two, of three or of any number of components.[1] The composition of the saturated solution changes suddenly when the nature of the solid phase or of the solution changes. The change of solvent cannot be shown at ordinary temperatures with the system, potassium chloride, potassium nitrate and water, because the salts do not act as solvent. In the system, potassium chloride, alcohol and water, we can pass from a solution with potassium chloride as solid phase and water as solvent to one with potassium chloride as solid phase and alcohol as solvent. In Fig. 39 is given the isotherm for sodium chloride, sodium nitrate and water at $20°$ taken from Nicol's[2] measurements and the isotherm for sodium chloride, alcohol and water at $30°$ taken from Bathrick's[3] determinations. The diagram is the one recommended by Schreinemakers, the co-ordinates being the concentrations of two of the components in a constant quantity of water, in this case grams of the salts and of alcohol in one hundred grams of water. The only change is that the logarithms of these concentrations are taken instead of the concentrations themselves. The abscissae are the logarithms of the concentrations of sodium chloride and the ordinates the logarithms of the concentrations of sodium nitrate and of alcohol. The scale for the alcohol is one-half that for the sodium nitrate.

The similarity of the two curves is very striking. Along AB sodium chloride is solid phase and along BC sodium nitrate, water being solvent along the whole length of the curve. There is a distinct break at B where the change occurs in the solute with respect

---

[1] Bancroft, Phys. Rev. **3**, 204 (1895); Jour Phys. Chem. **1**, 36 (1896).
[2] Phil. Mag. (5) **31**, 369 (1891).
[3] Jour. Phys. Chem. **1**. No. 3, 1896.

to which the solution is saturated. Along the whole of the curve A'B'C', sodium chloride is solid phase; but water is solvent along

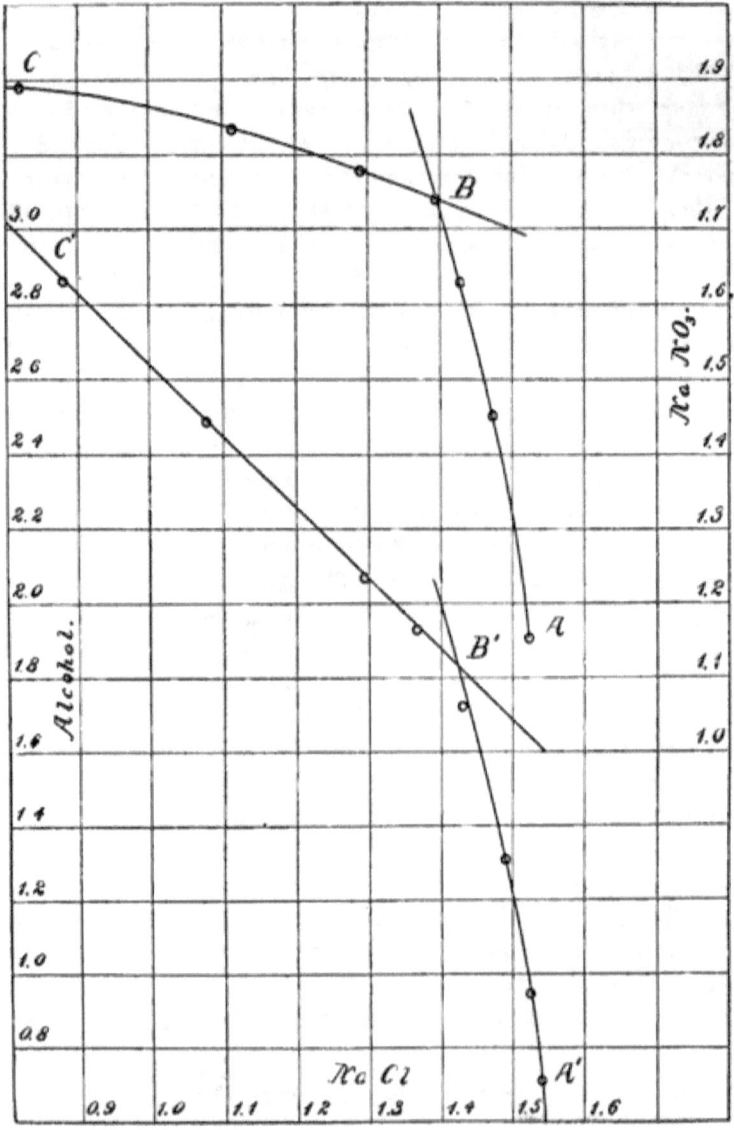

Fig. 39.

A'B' and alcohol along B'C'. There is a distinct break at B' where the nature of the solvent changes. The analogy between the two cases is complete.

Although no experimental work has yet been done illustrating the other cases which are represented in Fig. 36, it will be well to consider the general form of the isotherms in one or two particular instances as a guide to future investigators. For the sake of simplicity it will be assumed that the theorem of van Alkemade holds good, that the temperature rises along a boundary curve in the direction towards the line connecting the melting points of the two solid phases which exist along the boundary curve in question. There are so few exceptions to this rule that it is not worth while to consider them until they have been studied more fully, both experimentally and theoretically. In Fig. 40 are some of the cases from

Fig. 40.

Fig. 36 with a few characteristic isotherms sketched in. Ia is the diagram for a system forming one binary compound AC which is stable at its melting point, and which can coexist with any of the three components as solid phase. In the field DOKE the binary compound is solid phase. It is not known what is solvent in this field; but it is probable that the component A assumes that function in one part of the field and the component C over the rest of it. There seems no reason to assume that B is solvent anywhere outside of the field BGOKF. It should be clearly understood that this applies only to those cases where A and C are solvents in the respective fields in which they occur as solid phases. If A is solvent in part of the field in which C is solid phase, A is also solvent in the whole of the field occupied by AC. This case will occur with two salts and water, the only binary compound being a hydrate with a low inversion temperature.

At temperatures just below the melting point of the binary compound the isotherm will usually have the form represented by the lines marked 1, four isolated branches, one of them forming a semicircular curve round H, the melting point of the compound. This last branch may curve in at the bottom.[1] If the binary compound has a higher melting point than any of the components,[2] the first isotherm will consist of the curve round H and the other three branches will appear one by one as the temperature falls. At a lower temperature the isotherm marked 2 has become a closed curve made up of four parts. It is clear that the isotherm first becomes a closed curve at the lowest temperature at which any boundary curve meets the side of the triangle. This is a necessary geometrical consequence of this form of diagram and has no theoretical significance. The isotherm marked 3, is the first to meet the line OK. The general form is that of a figure eight. The contact occurs at the point where the line BH cuts OK and represents the maximum temperature at which AC and B can be in equilibrium together with the solution and vapor.

---

[1] Cf. Roozeboom, Zeit. phys. Chem. **15**, 602 (1894).

[2] The compound AuAl$_2$ has a higher melting point than pure gold so that a ternary system satisfying these conditions is possible. Cf. Roberts-Austen, Proc. Roy. Soc. **50**, 367 (1891).

If the temperature be raised at all, both solid phases will change into solution until one of them has disappeared. Which will disappear first depends on the relative quantities of the two solid phases. This maximum temperature has been called the fusion point of two solid phases.[1] It should be kept in mind that this intersection is a maximum temperature for the line OK but a minimum temperature for BH. At yet lower temperatures, the isotherm becomes two detached, closed curves contracting, one to the point O, the other to the point K. If the temperature of K is higher than that of O, the final result will be a single closed curve contracting to O as a vanishing point. If the temperature of K is lower than that of O the last isotherm possible will consist of a point at K.

In Ib the melting point H of the binary compound lies outside the field for $\overline{AC}$ and does not represent a state of stable equilibrium. Under these circumstances, the temperature will rise continuously as the system passes from O to K since this is in the direction towards the line BH. In this diagram the melting point of A is assumed to be lower than the temperature of the point K and the temperature of E to be lower than that of F. The first isotherm is for a temperature between those of the points E and F, the second for a temperature between K and E and the third for the temperature of the point K. Here for the first time, the part of the isotherm along which B is solid phase meets the part of the isotherm along which AC is solid phase. The point of intersection is no longer somewhere in the middle of the line OK nor is it on the line connecting B and H; but it is the highest temperature possible for the stable curve OK and the minimum temperature for the line BKH. If the appearance at K of C as solid phase can be prevented and the curve OK followed further it will be found that the temperature will continue to rise, passing through a maximum at the intersection of the prolongation of OK with DH. At lower temperatures the isotherm becomes a closed curve contracting to the point O.

In II the fields for the binary compound and for the third component are entirely separated.[2] The first isotherm is taken at the

[1] Roozeboom, Zeit. phys. Chem. **15**, 616 (1893).
[2] It is not at all certain that such a case can actually occur.

temperature of the point E to illustrate the fact that the temperature differences HE and HO are usually very unequal. The second isotherm in the diagram is a closed curve made up of four parts and the third a closed curve with three parts. This latter vanishes at the point O, because the temperature rises along the line OK according to the theorem of van Alkemade.

In IIIa there are two binary compounds represented by the conventional formulas $\overline{AC}$ and BC. Both have stable melting points and it is assumed that the line AR, if drawn, would cut OP at some point and that the line RH would cut PK. These two intersections will be maximum temperatures for OP and PK respectively. It is also assumed that the maximum temperature for OP is lower than the maximum temperature for PK, lower than the temperature of F and higher than the temperature of K. Under these circumstances the isotherm for the maximum temperature for OP has the general form of a figure eight round O and P and a detached closed curve round K as represented in the diagram. In IIIb the conditions are the same except that there is no maximum for OP, the temperature rising continuously from O to P. The isotherms in the diagram are for the maximum temperature for PK and for the temperature of the point P. In this diagram and the preceding one the final isotherm is the point O. In IIIc neither binary compound has a stable melting point and the melting point of A is assumed to be lower than the temperature of the points F and N. The first isotherm, marked 1, is that passing through the point K and the second the corresponding one for the point P. If the melting point R had been situated very close to the apex of the triangle the temperature would have risen in passing from P to O instead of falling as is now the case and the isotherm at P would have been merely a point.

When the curves for the monovariant systems do not pass through a maximum temperature, the parts of the isotherms in the adjoining fields first come in contact at the quintuple point terminating the curve at the higher temperature end. It has already been shown in discussing Ia that when there is a temperature maximum the two parts of the isotherm become tangent at that point. At lower temperatures they will meet at an angle.

CHAPTER XII

TWO SALTS AND WATER

In going over the systems which have been worked out experimentally, it will be best to begin with the one composed of magnesium sulfate, potassium sulfate and water, which has been studied by van der Heide[1]. Only a small portion of the entire field has been examined and this is reproduced in Fig. 41. The concentrations are

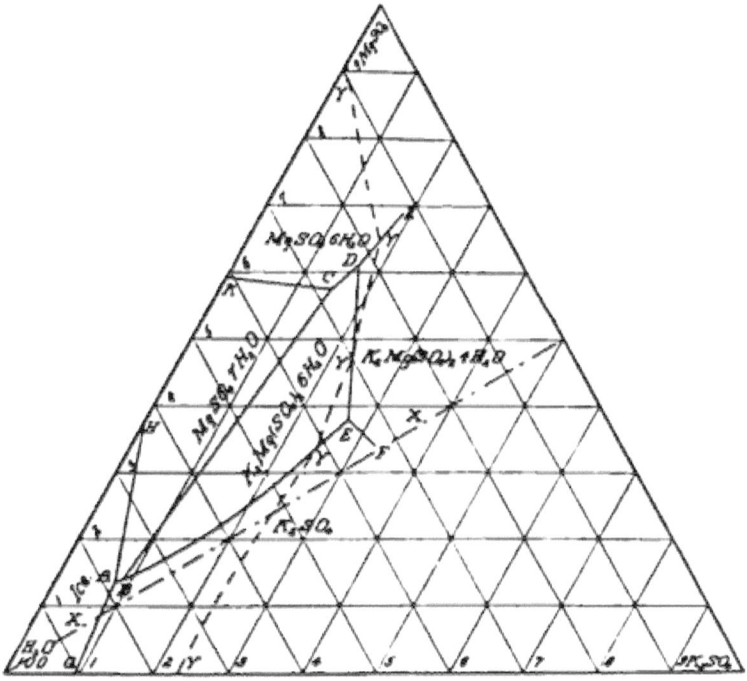

FIG. 41.

expressed in reacting weights, the sum of the three being always equal to one hundred. G is the cryohydric point for potassium sul-

---

[1] Zeit. phys. Chem. **12**, 416, (1893).

fate and ice, the temperature being $-1.2°$. H is the corresponding point for magnesium sulfate heptahydrate and ice, the temperature being $-6°$ owing to the greater solubility of the magnesium sulfate. Along GA ice and potassium sulfate are solid phases and along HA ice and magnesium sulfate heptahydrate. At A, $-4.5°$, there exists the nonvariant system, ice, magnesium sulfate with seven of water, potassium sulfate, solution and vapor. The temperature is higher than that of the point H, contrary to the rule of Schreinemakers[1] and van Alkemade. The reason for this is that potassium sulfate precipitates the heptahydrate of magnesium sulfate in such quantities that the solution actually becomes more dilute in passing from H to A, there being three and seven-tenths units of magnesium sulfate in solution at H while the sum of the two sulfates at A is only two and two-tenths units. Of course, potassium sulfate has a greater effect in lowering the vapor pressure of a solution than magnesium sulfate; but this is more than balanced by the difference in concentration so that the vapor pressure, and therefore the freezing point, of the solution at A is higher than at H. This is an interesting instance of the inapplicability of one of Ostwald's most important theorems. Ice, potassium sulfate and the heptahydrate of magnesium sulphate are in equilibrium at $-4.5°$ and therefore ice and the magnesium salt should be in stable equilibrium. As a matter of fact, unless there is an error in the determinations, a mixture of these two substances will liquefy at this temperature.

Along the curve B the solid phases are the two sulfates. At B, $-3°$, the two sulfates unite to form a ternary compound having the formula $K_2Mg(SO_4)_2, 6H_2O$. We have thus at B, a new nonvariant system, with potassium sulfate, magnesium sulfate heptahydrate and the hydrated double salt as solid phases. Along BC the solid phases are the hydrated double salt and magnesium sulfate heptahydrate. At C, $47.2°$, the salt with seven of water passes into that with six of water and there is formed another nonvariant system with the heptahydrate, the hexahydrate and hydrated double salt as solid phases. Along CK the solid phases are the heptahydrate and the hexahydrate of magnesium sulfate; K, $48.2°$, represents the temperature

[1] Zeit. phys. Chem. **12**, 77, (1893).

and concentration at which magnesium sulfate with seven and six of water are in equilibrium with solution and vapor when no potassium sulfate is present. In other words it is the inversion point for the two hydrates in the binary system, magnesium sulfate and water. Along CD the solid phases are the hydrated double salt and magnesium sulfate with six of water.[1] At D, 72°, another solid phase appears in the form of a second hydrated double salt with the formula $K_2Mg(SO_4)_2, 4H_2O$. Along DL the solid phases are the new hydrated double salt and the hexahydrate of magnesium sulfate. This curve has not been followed very far; but an experiment in a sealed tube showed that at a temperature of 106° a change takes place presumably of the hexahydrate into a monohydrate. Along DE the solid phases are the two hydrated double salts. At E, 92°, the solution becomes saturated in respect to potassium sulfate forming yet another nonvariant system. Along EF the solid phases are potassium sulfate and the salt $K_2Mg(SO_4)_2, 4H_2O$. At some point on this curve the hydrated double salt will lose water changing into something else; but it is not known what this change is nor at what temperature it takes place. Along EB the solid phases are potassium sulfate and the hydrated double salt $K_2Mg(SO_4)_2, 6H_2O$. In the field to the left of HAC ice is the solid phase; in the field HABCK magnesium sulfate with seven of water. The hexahydrate exists in the field bounded by KCDL and the undetermined line for the monohydrate. Potassium sulfate occurs as solid phase in the field bounded by GABEF and other lines not yet determined. The field for $K_2Mg(SO_4)_2, 6H_2O$ is entirely determined, being contained within the closed figure BCDEB. On the other hand, the field for the other hydrated double salt $K_2Mg(SO_4)_2, 4H_2O$ is bounded on the left by the lines LDEF while the right hand boundary is unknown.

If we draw the dotted line XX through the lower left hand corner of the triangle and the middle of the opposite side, this line represents a series of solutions in which potassium and magnesium sulfates are present in equivalent quantities. A point P on this line

---

[1] There is a second modification of magnesium sulfate hexahydrate; but it is labile at all temperatures and is not considered here. Cf. van der Heide, Zeit. phys. Chem **12**, 421 (1893).

which does not happen to fall within the limits of the present diagram would represent a solution having the same composition as the hydrated double salt with six of water while yet another point Q farther out would represent a solution having the same composition as the second hydrated double salt, the one with four of water. Whether this second point lies within the field for the salt K_2MgSO_4 $4H_2O$ is not known because the boundaries of this field have not yet been fully determined; but it is improbable that this is the case. We know, however, that the solution having a composition represented by the formula $K_2Mg(SO_4)_2 6H_2O$ does not lie within the field in which that compound can exist as a solid phase. In other words this particular hydrated double salt can not exist in equilibrium with a solution having the same composition as itself and therefore has not a true melting point. From the position of the dotted line XX there are other conclusions to be drawn. It does not cut the field for $K_2Mg(SO_4)_2 6H_2O$ at any point and therefore this salt can not be in equilibrium with any solution in which potassium sulfate and magnesium sulfate are present in equivalent quantities. Addition of water to this salt will therefore decompose it, dissolving out an excess of one component and leaving some of the other as solid phase.[1] In this particular case addition of water to the hydrated double salt at any temperature between $-3°$ and $92°$ will cause a partial decomposition with formation of potassium sulfate as solid phase, the solution having the concentration corresponding to some point on the curve BE. Further addition of water will cause increased formation of potassium sulfate and an increased amount of solution, the concentration remaining constant until the whole of the hydrated double salt has disappeared, leaving the divariant system, potassium sulfate, solution and vapor. On adding more water the potassium sulfate will dissolve giving finally a series of unsaturated solutions in which the potassium and magnesium sulfates are at last present in equivalent quantities. With the hydrated double salt, $K_2Mg(SO_4)_2 4H_2O$, the case may be different. If the dotted line XX cuts the line EF, as it seems to in the diagram, this ternary compound will not be de-

---

[1] Cf. Roozeboom, Zeit. phys. Chem. **2**, 520 (1888); Schreinemakers, Ibid. **9**, 75 (1892).

composed by water. Unfortunately the line EF has only been followed a very short way and it may come to an end before it reaches the point F in the diagram. All that can be said is that if the dotted line cuts the field for the hydrated double salt, $K_2Mg(SO_4)_2, 4H_2O$, this compound will not be decomposed by water; otherwise it will. It is not of much importance either way as there are cases known where the ternary compound is not decomposed by water.

In regard to the changes of direction at the inversion points, there is a theorem by Meyerhoffer[1] which has a certain qualitative value. It is that the solubility curve for the solid phase which does not disappear has no break at the inversion temperature. While it is necessarily true that the concentration curve for the component which disappears as solid phase will always have a break at the inversion point, the converse of this, which is the theorem of Meyerhoffer, will never be true exactly though the approximation may be very close. A change in the nature of the second solid phase will necessarily affect the solubility of the first because it is impossible that two different substances can cause the same precipitation, positive or negative, of another body over a range of temperatures. On the other hand, this difference of effect will be relatively small and can usually be ignored. The great difficulty in regard to this theorem of Meyerhoffer is that the way in which it is deduced would seem to imply that the concentrations should be given as amounts of each salt in a constant quantity of water or in a constant quantity of water plus that salt whereas Meyerhoffer takes the amounts of each salt in one hundred grams of the solution containing water and both salts. He is not consistent even in this because the data of Roozeboom[2] are advanced as proofs of the theorem, though these are expressed as reacting weights of each salt in one hundred reacting weights of water.

This theorem can be applied in the system under discussion. Along the line BCD potassium magnesium sulfate with six of water is one of the solid phases, magnesium sulfate with seven of water being the other solid phase along BC and magnesium sulfate with

---

[1] Zeit. phys. Chem. **5**, 120 (1890).
[2] Ibid. **2**, 518 (1888).

six of water along CD. It is found that the amount of potassium sulfate in solution changes practically continuously with the temperature while the amount of magnesium sulfate in the unit quantity of solution changes discontinuously as the temperature becomes 47.2°, at the point C. Along the curve CDL the concentration of potassium sulfate changes discontinuously at D while that of magnesium sulfate does not. Along BEF the reverse is the case. On the other hand, Meyerhoffer seems to have found, in the system, sodium and magnesium sulfates, and water, a case where both curves show a break. This however is not well established.[1]

With the triangular diagram one can not tell whether Meyerhoffer's theorem holds good unless the isotherms are marked; but this is shown directly in the diagram of van der Heide which is really a double diagram.[2] The temperature is taken as one axis and the concentrations of the two salts in one hundred units of solution are laid off on the other axis, right and left from a zero point. In this method there are two points for each nonvariant system and two curves for each monovariant system. This working in duplicate is a disadvantage; but it is, perhaps, compensated by having the temperature as an axis in the plane of the paper.

The direction of the temperature changes along the different curves can be predicted, with a single exception, from the theorem of van Alkemade. As will be remembered, this theorem states that the temperature rises when going along a boundary curve in the direction towards the line connecting the melting points of the two solid phases in the system. The exception is the line AH. The line connecting the melting points is the left side of the triangle and the temperature of the point H should be higher than that of A; but this is not the case. For AC, AB, BC, CK, CD, DL, DE, EF and BD the theorem applies. It is to be noticed that XX is the line connecting the melting points of the two hydrated double salts so that it is entirely proper that the temperature of the point E should be higher than that of the point D.

Only one isotherm is represented, that for 85°. The line YY

---

[1] Zeit. phys. Chem. 5, 122 (1899).
[2] Ibid. 12, 425 (1893).

shows its course, but merely approximately as there are not sufficient data to permit of its being drawn accurately to scale. It is composed of four parts while the isotherm for 72° and 92° have three parts only. At temperatures between 47.2° and 48.2° the isotherm will be made up of four sections because it cuts the lines KC, CD and EB. Since the lowest temperature at which a solution can exist is the point H in the side of the triangle, there will be no isotherm which will form a closed curve. At −4.5° the isotherm will consist of two lines from the point A meeting the side of the triangle, one above and the other below the point H. With falling temperature the intersection of the two lines will pass along the curve AH until at H the isotherm becomes a point. The data for the system composed of magnesium sulfate, potassium sulfate and water are given in Table XXVI. The concentrations are reacting weights in one hundred reacting weights of solution, x referring to potassium sulfate and y to magnesium sulfate.

Table XXVI

Temp.	x	y	Temp.	x	y
Curve BE			Curve AB		
− 3.°	0.89	1.37	− 4.5°	0.87	1.34
+10.	1.17	1.63	Curve BC		
20.	1.39	1.84	+22.	1.24	3.78
30.	1.63	2.07	47.3	1.53	5.75
40.	1.87	2.37	Curve CD		
50.	2.05	2.76	72.	1.69	6.12
60.	2.17	3.11	Curve DL		
70.	2.25	3.24	85.	1.78	6.46
80.	2.47	3.60	Curve EF		
90.	2.58	3.65	98.	2.73	3.78
92.	2.67	3.79			

In the system just studied, one of the hydrated double salts was certainly decomposed by water at all temperatures and the other may have been. In the system consisting of sodium sulfate, magnesium sulfate and water, there is formed a hydrated double salt which is not decomposed by water at certain temperatures. This equilibrium has been studied to a certain extent both by van 't Hoff[1] and by Roozeboom.[2] The solubility determinations are represented graphically in Fig. 42, the concentrations being reacting weights in one

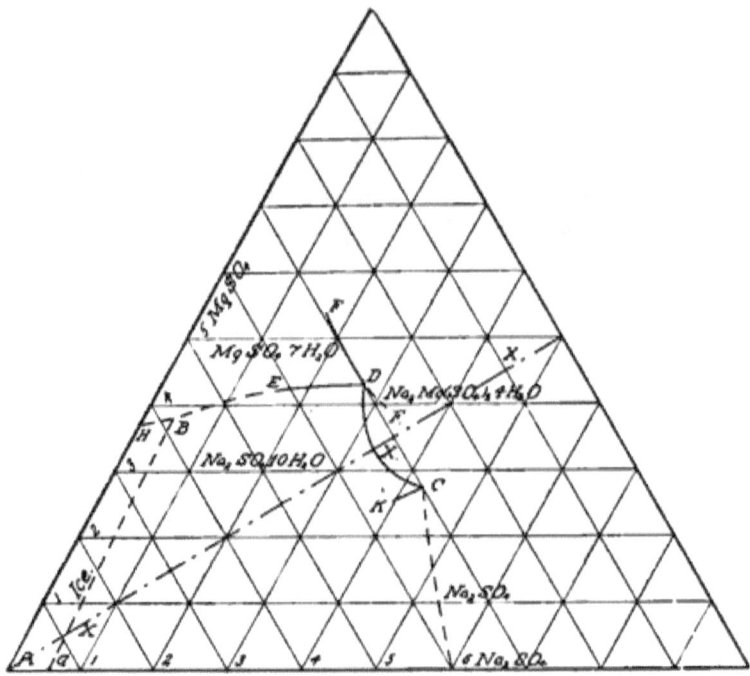

FIG. 42.

hundred reacting weights of the solution. H and G are the cryohydric points for magnesium sulfate heptahydrate and sodium sulfate decahydrate respectively, the temperatures being $-6°$ and $-0.7°$. At B the solid phases are the two single salts and ice. The temperature at which this nonvariant system exists is unknown as well as the concentration of the solution. For this reason the three curves meeting at this point are represented by dotted lines. Along BD the solid phases are the two hydrated salts. No points on this curve have been determined below E, $15°$. At D, $21.5°$ there appears the

[1] Zeit. phys. Chem. **1**, 165 (1887).
[2] Ibid. **2**, 518 (1888).

hydrated double salt corresponding to the formula $Na_2Mg(SO_4)_2$, $4H_2O$. Along FPF_1 the solid phases are magnesium sulfate heptahydrate and the hydrated double salt. In the direction PF this curve will be terminated by the appearance of the hexahydrate as solid phase. The co-ordinates of this new nonvariant system have not been determined. The part of the curve PF_1 represents a labile equilibrium, instable with respect to the decahydrate of sodium sulfate. Along the curve DC the solid phases are the hydrated double salt and hydrated sodium sulfate. At C, 30°, there appears a new solid phase, namely anhydrous sodium sulfate.[1] Along CK the solid phases are anhydrous sodium sulfate and the hydrated double salt. This curve has only been followed a little way and it is not known what becomes of it at higher temperatures. It is rather important that this should be investigated since it is impossible to predict whether the curve will bend upwards or downwards. Its present direction is contrary to the theorem of van Alkemade ; but we are not justified in calling it a real exception until the whole curve has been determined. From the point C there must run a third curve along which anhydrous and hydrated sodium sulfate are solid phases ; but the only other point on the curve which is known is at N, 32.6°, for the binary system composed of sodium sulfate and water. For this reason a dotted line has been drawn between these two points to emphasize the fact that they are connected. The experimental curve will probably have this general form.

The fields for the divariant systems, in which there is only one solid phase, are marked on the diagram so far as possible. Ice exists in the field bounded by GBH ; magnesium sulfate heptahydrate in the field bounded by HBDF and the undetermined line for the hexahydrate. The field for sodium sulfate with ten of water is bounded by GBDCN. The field for anhydrous sodium sulfate can not be given until something definite is known about the curve CK. It will be possible to have two different divariant systems in the field between CK and CN and also in the field between CK and CD.

---

[1] The observation of van 't Hoff that the hydrated sodium sulfate melts at 26° in presence of an equivalent quantity of crystallized magnesium sulfate can not refer to a state of stable equilibrium and has no place in this discussion.

In the field KCN one of the solid phases will be hydrated sodium sulfate while the solid phase of the second divariant system will be anhydrous sodium sulfate or the hydrated double salt according as CK curves in one direction or the other. Although no measurements have been made in regard to this, Roozeboom feels certain that anhydrous sodium sulfate exists as the solid phase to the right of CK and the hydrated double salt to the left.[1] It is, of course, clear that two sets of divariant systems can not exist at the same temperature with the same concentrations in the solution phase. We can, therefore, have different isotherms apparently intersecting in the fields KCN and KCD. This indefiniteness disappears if we consider the solid figure formed with the temperature as vertical axis. The reason for the complication is that the solubility of anhydrous sodium sulfate decreases with increasing temperature.

While we do not know the whole of the field for the hydrated double salt, it is clear from the diagram that this compound can exist as solid phase in the field to the right of the lines FDC. The dotted lines XX represent solutions which contain the two salts in equivalent quantities. This line cuts the curve DC at the point X_1 corresponding to a temperature of $25°$, showing that the hydrated double salt is decomposed by water at a temperature below $25°$ while, from this temperature on, water will dissolve the solid components in equivalent quantities. This is found to be the case experimentally. Above $25°$ the hydrated double salt is not decomposed by water; but dissolves and recrystallizes as if it were a single substance. Below $21.5°$ the double salt cannot exist at all and between $21.5°$ and $25°$ it can exist only in presence of a solution containing a relatively larger amount of magnesium sulfate than the crystals. In other words between these two temperatures the hydrated double salt is decomposed by water with precipitation of hydrated sodium sulfate. The temperatures between the limits within which the double salt is decomposed have been called the range of decomposition by Meyerhoffer.[2] For this particular compound the range is only about three and a half degrees; for the double sulfate

---

[1] Private letter.
[2] Zeit. phys. Chem. 5, 109 (1890).

of magnesium and potassium with six of water we have seen that it covered all the temperatures at which that salt could exist. With the double sulfate of copper and potassium the range of decomposition is presumably zero though there are no experiments to that effect. Within the range of decomposition a compound can not be purified by recrystallization because it is continually decomposed by water. In Table XXVII are the data for the system, magnesium and sodium sulfates and water. The concentrations are expressed as reacting weights of the salts in one hundred reacting weights of the solution, x denoting sodium sulfate and y magnesium sulfate.

TABLE XXVII

Temp.	x	y	Temp.	x	y
Curve BED			Curve F,DF		
15.°	1.55	4.25	18.5°	3.17	3.96
18.5	2.01	4.29	22.	2.66	4.32
Curve DC			24.5	2.51	4.44
22.	2.74	4.37	30.	2.14	4.93
24.5	3.23	3.39	35.	1.61	5.46
30.	4.27	2.71	Curve CK		
			35.	4.02	2.58

The equilibrium between **cupric chloride, potassium chloride and water has been studied by Meyerhoffer**[1] though his investigations cover only a small portion **of the field. His results are represented graphically in Fig. 43. G is the cryohydric point for ice and potassium chloride, the temperature being** $-11.4°$ **according to Guthrie**[2]. H is **the corresponding point for cupric chloride with two of water. This point has not been accurately determined but the temperature is lower than** $-23°$ [3]. **Along GB ice and potassium chloride are solid phases and along HN ice and hydrated cupric chloride. These curves do not meet, as in all probability the double chloride with the composition corresponding to the formula CuCl₂2KCl2H₂O appears both at B and at N. Each of these points represents a nonvariant**

---

[1] Zeit. phys. Chem. **3**, 336 (1889).
[2] Phil. Mag. (4) **49**, 269 (1875).
[3] Cf. de Coppet. Ann. chim. phys. (4) **23**, 386 (1871).

system, the solid phases at B being ice, potassium chloride and copper dipotassium chloride, while at N they are ice, hydrated cupric chloride and the hydrated double salt. Along BN the solid phases are ice and the hydrated double salt. Not one of these curves has been studied experimentally; but the temperature should rise in passing from N to B. Along BD the solid phases are potassium chloride and the blue copper dipotassium chloride. This curve has been followed from E, 39°, to D, 92°. At this latter temperature

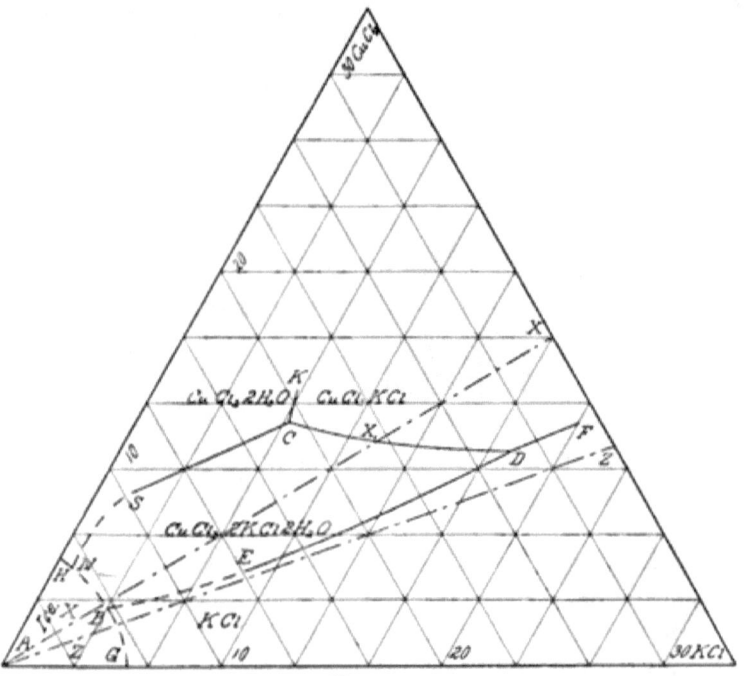

FIG. 43.

there is formation of the red anhydrous double salt represented by the formula $CuCl_2.KCl$. At D there exists the nonvariant system, potassium chloride, copper dipotassium chloride, copper potassium chloride, solution and vapor. If the temperature rises above 92° the copper dipotassium chloride breaks up into potassium chloride and copper potassium chloride forming the monovariant system stable along DF, the two last mentioned salts being present as solid phases. Going back and starting from N we can pass along the curve NC with hydrated cupric chloride and the blue ternary compound as solid phases. The point S corresponds to the temperature of zero degrees centigrade and is the point at which Meyerhoffer's measurements begin. Beyond C, 56°, the copper dipotassium chloride ceases to be stable in presence of an excess of cupric chloride and, at this point, there exists the nonvariant system, hydrated cupric chloride, copper dipotassium chloride, copper potassium chloride, solution and vapor. Along CK the solid phases are hydrated cupric chloride and the anhydrous double salt while along CD they are the ternary and the binary compounds of copper and potassium chloride.

Ice exists as solid phase in the field bounded by HNBG, potassium chloride in that bounded by GBDF; and hydrated cupric chloride in the field bounded by HNCK and the undetermined curve for the anhydrous salt. The blue hydrated double salt exists as solid phase only in the closed field DBNCD. The dotted line ZZ represents solutions in which there is twice as much potassium chloride as cupric chloride. This line does not cut the field for copper dipotassium chloride so far as we know and therefore this salt is always decomposed by water with precipitation of potassium chloride. The green color which this salt usually has is due to a slight decomposition and traces of mother liquor rich in cupric chloride. To obtain the salt pure it should be washed with a potassium chloride solution instead of with pure water. The dotted line XX represents solutions in which the ratio of cupric and potassium chlorides is unity. The field for the anhydrous double salt, bounded by KCDF, is cut by this line at N, 72°, and from this temperature upwards it is possible to obtain a solution of this double salt containing the two salts in the same ratio as in the solid phase. Between 56° and 72° the red double salt is decomposed by water with formation of the blue salt $CuCl_2 \cdot 2KCl \cdot 2H_2O$. In this case the range of decomposition extends over sixteen degrees.

The line ZZ connects the melting points of ice and of copper dipotassium chloride and it is for this reason that the temperature should rise as the system passes along NB from N to B. It is to be noticed that the salt component which is present at B is necessarily the component which is precipitated by the action of water upon the double salt. This has been expressed in another form by Schreinemakers[1] in the rule: "The cryohydric temperature of a solution in equilibrium with double salt and the component which does not precipitate is lower than the cryohydric temperature of a solution in equilibrium with double salt and the component which precipitates." If the double salt were not decomposed by water, the points N and B in the diagram would lie on opposite sides of the line ZZ and there would be, according to the theorem of van Alkemade, a maximum temperature at the point where ZZ cuts NB. This conclusion was drawn simultaneously by Schreinemakers[2] and by Meyerhoffer,[3] who reached this result independently and by different ways. As an illustration of this point we may take the system, copper sulfate, ammonium sulfate and water. The temperature at which ice, copper sulfate with five of water and the hydrated double salt $Cu(NH_4)_2(SO_4)_2,6H_2O$ are in equilibrium with solution and vapor is $-2.6°$. The temperature for the corresponding point with ammonium sulfate instead of copper sulfate as solid phase is $-19°$. A solution containing the two salts in equivalent quantities is in equilibrium with ice and the double salt at $-1.7°$. The double salt of copper and ammonium chloride $CuCl_2,2NH_4Cl_2H_2O$ is in equilibrium with ice and ammonium chloride at $-15.7°$; with ice and hydrated cupric chloride at a temperature lower than $-42°$, while the curve connecting these points passes through a maximum temperature at $-12.7°$.

In table XXVIII are the data for the system, potassium chloride, cupric chloride and water, x denoting reacting weights of potassium chloride, y reacting weights of cupric chloride in one hundred reacting weights of solution.

The system, lead iodide, potassium iodide and water studied by Schreinemakers[4] need not detain us long. There is formed a com-

---

[1] Zeit. phys. Chem. **12**, 85 (1893).
[2] Ibid. **12**, 87 (1893).
[3] Sitzungsber. Akad. Wiss. Wien, : **02**, IIb, 200 (1893).
[4] Zeit. phys. Chem. **9**, 57; **10**, 467 (1892); Schreinemakers was misled by a faulty analysis made by Ditte and, in the first paper, the formula of the hydrated double salt is wrong and consequently most of the theoretical conclusions. In the second paper the errors are corrected.

pound, PbI$_2$KI2H$_2$O, which is decomposed at all temperatures with precipitation of **hydrated lead iodide.** The cryohydric temperature for double **salt and lead iodide is —2.8°** and for double salt and potassium iodide is **—22.8°. This is an** experimental confirmation of the rule of Schreinemakers **in regard** to the relative temperatures **of these points.** The **hydrated double salt is an** excellent one for a lecture experiment since **it is pale yellow in color, while the hydrated lead iodide crystallizes in lustrous almost orange scales. Addition of water produce an almost instantaneous decomposition with change of color. This can be made to disappear by adding potassium iodide.**

TABLE XXVIII

Temp.	x	y	Temp.	x	y
Curve BED			Curve NSC		
39.4°	8.61	4.82	0.°	1.7	8.81
49.9	9.68	5.43	39.6	4.6	10.9
60.4	11.2	6.35	50.1	5.7	11.38
79.1	14.47	8.54	52.9	6.25	12.08
90.5	17.5	10.69	Curve CK		
Curve DF			60.2	6.84	12.74
93.7	18.3	11.4	72.6	6.68	13.43
98.8	20.1	12.3	Curve CD		
			64.2	9.16	11.78

There are no other ternary **systems made up of two salts and** water which have been studied **at all in detail ; but there are determinations of** several quintuple points **or points at which five phases coexist.**[1] Calcium acetate **combines with copper acetate to form a double salt**[2] which **splits into the component salts at 76°. The solid phases at this point are** CaAc$_2$H$_2$O, **CuAc$_2$H$_2$O and CaCuAc$_4$8H$_2$O. This is a** striking experiment **because copper acetate is green and calcium** acetate white, while the **hydrated double salt is an intense blue.** In this case the double salt **is stable below the inversion temperature** while the **double sulfate of magnesium and sodium was** stable at higher temperatures, **decomposing into single salts at tem-**

---

[1] Roozeboom, Recueil Trav. Pays-Bas **6**, 331, (1887) ; **Zeit. phys. Chem. 2,** 513, (1888).

[2] Reicher, **Ibid. 1,** 221, (1887).

peratures below that of the quintuple point. The Theorem of Le Chatelier would tell us the direction of the change at the quintuple point if the heats of reaction were known. As this is generally not the case, we must content ourselves with an approximation. If we assume that no one of the compounds has a true melting point, the following rule holds good in all the instances yet studied. When one of the solid phases can split into the other two with addition or subtraction of water[1] the inversion point is a minimum temperature for that phase if the water be added to complete the reaction and a maximum temperature if the water be subtracted.[2]

The following instances will illustrate this rule and at the same time furnish a list of the different systems in which quintuple points have been determined. At $-3°$ one of the double sulfates of magnesium and potassium changes into the single sulfates with addition of water.[3] This is a minimum temperature for $K_2Mg(SO_4)_2 6H_2O$.

$$K_2Mg(SO_4)_2 6H_2O + H_2O = K_2SO_4 + MgSO_4 7H_2O.$$

The same change takes place at $21.5°$ with the double sulfate of sodium and magnesium.[4] This is the minimum temperature for the double sulfate, $Na_2Mg(SO_4)_2 4H_2O$.

$$Na_2Mg(SO_4)_2 4H_2O + 13H_2O = Na_2SO_4 10H_2O + MgSO_4 7H_2O.$$

Copper potassium chloride changes at $56°$ into copper dipotassium chloride and hydrated cupric chloride.[5] This is a minimum temperature for $CuCl_2 KCl$.

$$2CuCl_2 KCl + 4H_2O = CuCl_2 2KCl 2H_2O + CuCl_2 2H_2O.$$

At $92°$ copper dipotassium chloride changes into copper potassium chloride and potassium chloride. This is a maximum temperature for $CuCl_2 2KCl 2H_2O$.

$$CuCl_2 2KCl 2H_2O - 2H_2O = CuCl_2 KCl + KCl.$$

---

[1] It is not possible to have in equilibrium three solid phases such that one can be formed from the other two without addition or subtraction of water.

[2] Cf. Roozeboom, Recueil Trav.-Pays-Bas, **6**, 341 (1887); Zeit. phys. Chem. **2**, 517 (1888), Bancroft, Jour. Phys. Chem. **1**, No. 6 (1897).

[3] van der Heide, Zeit. phys. Chem. **12**, 416 (1893).

[4] van 't Hoff and van Deventer, Ibid. **1**, 165 (1887).

[5] Meyerhoffer, Ibid. **3**, 336 (1889).

An analogous change occurs at 146° with copper diammonium chloride.[1] This is a maximum temperature for $CuCl_2,2NH_4Cl,2H_2O$.

$$CuCl_2,2NH_4Cl,2H_2O - 2H_2O = CuCl,NH_4Cl + NH_4Cl.$$

At 15.5° one of the double salts of copper tetrethyl ammonium chloride changes into another double salt and hydrated cupric chloride.[2] This is a minimum temperature for $5CuCl_2,2N(C_2H_5)_4Cl$.

$$5CuCl_2,2N(C_2H_5)_4Cl + 8H_2O = CuCl_2,2N(C_2H_5)_4Cl + 4CuCl_2,2H_2O.$$

The double acetate of copper and calcium changes at 76° into the single acetates.[3] This is a maximum temperature for $CuCaAc_4,8H_2O$.

$$CuCaAc_4,8H_2O - 6H_2O = CuAc_2,H_2O + CaAc_2,H_2O.$$

Sodium ammonium racemate decomposes at 27° into the dextrorotary and laevorotary sodium ammonium tartrates.[4] This is a minimum temperature for $(NaNH_4C_4H_4O_6,H_2O)_2$.

$$(NaNH_4C_4H_4O_6,H_2O)_2 + 6H_2O = 2(NaNH_4C_4H_4O_6,4H_2O).$$

The same salt changes at 35° into the single racemates.[5] This is a maximum temperature for $(NaNH_4C_4H_4O_6,H_2O)_2$.

$$2(NaNH_4C_4H_4O_6,H_2O)_2 - 4H_2O = (Na_2C_4H_4O_6)_2 + ([NH_4]_2C_4H_4O_6)_2.$$

The double potassium sodium racemate undergoes similar changes at the temperature of −6° and 41° respectively[6] the first being a minimum temperature for $(KNaC_4H_4O_6,3H_2O)_2$, and the second a maximum temperature for the same salt.

$$(KNaC_4H_4O_6,3H_2O)_2 + 2H_2O = 2(KNaC_4H_4O_6,4H_2O).$$

$$2(KNaC_4H_4O_6,3H_2O)_2 - 8H_2O = (Na_2C_4H_4O_6)_2 + K_2C_4H_4O_6,2H_2O)_2.$$

There are some nonvariant systems containing three solid phases, solution and vapor, in which two of the solid phases can not be made from the third with or without addition of water and these must also be considered. They can be divided into two classes: "One solid phase can be transformed into one of the others by ad-

---

[1] Meyerhoffer, Zeit. phys. Chem. **5**, 98 (1890).
[2] Meyerhoffer, Sitzungsber. Akad. Wiss. Wien. **102**, IIb, 150 (1893).
[3] Reicher, Zeit. phys. Chem. **1**, 221 (1887).
[4] van 't Hoff and van Deventer, Ibid. **1**, 165 (1887).
[5] van 't Hoff, Goldschmidt and Jorissen, Ibid. **17**, 49 (1895).
[6] van 't Hoff and Goldschmidt, Ibid. **17**, 505 (1895).

dition or subtraction of water. No one of the solid phases can be converted into either of the others by addition or subtraction of water." If one solid phase can be converted into one of the others by addition of water the inversion point is a maximum or a minimum temperature for one of those phases and is neither a maximum nor minimum for the the third phase. This can be illustrated very readily by three instances from the system, potassium sulfate, magnesium sulfate and water. At 47.2° two of the solid phases are $MgSO_4, 7H_2O$ and $MgSO_4, 6H_2O$ while the third is $K_2Mg(SO_4)_2, 6H_2O$. This is a minimum temperature for the hexahydrate. If the temature of the point had been higher than 48.2° it would have been a maximum temperature for the heptahydrate. No example of this latter form has yet been observed; but it is not impossible. The hydrated double salt exists both above and below the temperature of the inversion point. At 72° two of the solid phases are the hydrated double salts $K_2Mg(SO_4)_2, 6H_2O$ and $K_2Mg(SO_4)_2, 4H_2O$ while the third is magnesium sulfate heptahydrate. At 92° two of the solid phases are the same two hydrated double salts and the third is potassium sulfate. The first temperature is a minimum for the double salt with four of water and the second a maximum for that with six of water. If the temperatures are not given it can only be told by experiment which point is which. If the compositions of the solutions are known, the direction of the temperature change can be foretold from the theorem of van Alkemade. The higher temperature will necessarily be a maximum for the double salt with the larger amount of water of crystallization.

When no one of the three solid phases can be converted into either of the others, it is impossible to make any definite prediction when the only data are the formulas of the three solid phases. As an instance take the two quintuple points where the solid phases are ice, hydrated calcium acetate and copper calcium acetate, ice, hydrated copper acetate and copper calcium acetate. The two sets consist of ice, a hydrated salt and a hydrated double salt. There is no way of distinguishing them without further information. Here again the theorem of van Alkemade will help us if the concentrations of the two solutions are known and if the double salt is decomposed by water. If the double salt is stable in presence of water, there is no *a priori* method of telling which cryohydric tempreature is the higher, though this may be guessed at if the concentrations are known.[1] It should be clearly understood that in all these cases a maximum or a minimum temperature for a given substance refers to that substance in equilibrium with solution and vapor. For instance, $56°$ is a minimum temperature for copper potassium chloride in equilibrium with solution and vapor; but it is possible for copper dipotassium chloride, copper potassium chloride, potassium chloride and vapor to be in stable equilibrium at ordinary temperatures.

It is not an entirely empirical rule that when one solid phase can change into the other two with addition or subtraction of water the quintuple point is a minimum or a maximum temperature respectively for that phase. The heat effect due to the absorption or splitting off of water is so much greater than any of the other heats of reaction that it determines the sign of the whole heat effect and the direction of the change. In cases where this is not so, the rule will not hold. One would expect it to apply in practically all cases where two of the components are solids in the pure state at all temperatures covered by the experiment, and where the third component is near or above its melting point. It so happens that most systems which have been studied come under this head. Water occupies an exceptional position in determining the direction of the change because it occupies an exceptional place in the system, being the solvent and practically the only constituent in the vapor phase. It is obvious that it is necessary to limit these rules for the change of the equilibrium with the temperature to compounds not having a true melting point, because the melting point is always the maximum temperature for that phase, and neither of the two quintuple points in which a boundary curve terminates is necessarily a maximum temperature even for the monovariant system which exists along it.

[1] Cf. Schreinemakers, Zeit. phys. Chem. **12**, 73 (1893).

CHAPTER XIII

PRESSURE CURVES

All the monovariant systems considered thus far have been composed of two solid phases, solution and vapor, for only these find a place in a concentration-temperature diagram. In the quintuple point five boundary curves meet so that there are still two monovariant systems unaccounted for. Instead of taking these by themselves, it will be better to treat them as parts of the pressure-temperature diagram for a system of three components, only one of which is measurably volatile. In the simplest case that we need consider, **one of** the salts forms a hydrate, which can coexist **as** solid phase **with any** of the three other possible solid phases, and is instable at **its melting** point. The concentration-temperature diagram for this system **is** given by I a in Fig. 36. Such a system would be realized **with sodium** sulfate, sodium chloride and water, denoted by C, B and **A, respectively.** Fig. 44 is the pressure-temperature diagram for **the equi**libria around the quintuple point K where sodium chloride, **hydrated** and anhydrous sodium sulfate, solution and **vapor coexist. KF, KE**

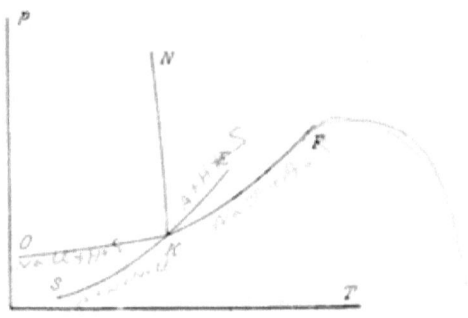

FIG. 44.

and KO are the curves for the monovariant systems composed of two solid phases, solution and vapor. The solid phases are hydrated and anhydrous sodium sulfate along KE, anhydrous sodium sulfate and sodium chloride along KF, hydrated sodium sulfate and sodium chloride along KO. At the point E the concentration of sodium chloride is zero, and the concentration of water is zero at the point F. The curve KO terminates at the point O where ice appears as a solid phase, making a new quintuple point. The solution connoted by KE is more dilute than the one connoted by KF and therefore the former curve lies above the latter. The curves KE and KF, if prolonged, will lie below KO. When one of the phases is a saturated solution, the more stable system is the one with the higher vapor pressure.[1] This is true for ternary as well as for binary systems. It does not follow from this that the monovariant system represented by KF is instable with respect to that represented by KE. It is not sufficient to consider the total concentrations only; with three components the relative amounts of the two salts is a very important factor in determining the equilibrium. If the solution with the lower vapor pressures along KF could change into the solutions with the higher vapor pressures along KE by precipitation of hydrated sodium sulfate, the first set of solutions would not represent stable modifications. This is not the case, and both KF and KE are pressure-temperature curves for stable monovariant systems.

Of the other two curves meeting in the point K, the curve KN represents the conditions of pressure and temperature under which hydrated and anhydrous sodium sulfate, sodium chloride and solution are in equilibrium. According to the theorem of Le Chatelier, this curve will slant to the left if hydrated sodium sulfate and sodium chloride occupy a larger volume than anhydrous sodium sulfate and the resulting solution; to the right if the reverse is the case. The curve KS represents the equilibrium between hydrated and anhydrous sodium sulfate, sodium chloride and vapor. This curve, if prolonged, would certainly lie above KF and possibly above KE because the system with the higher vapor pressure is the less stable when it contains no solution phase.

The quantitative relations of EK and KS to the boundary curves for the binary system, sodium sulfate and water, are not definitely known. In Fig. 45 are given the two chief possibilities. The quad-

---

[1] See page 58.

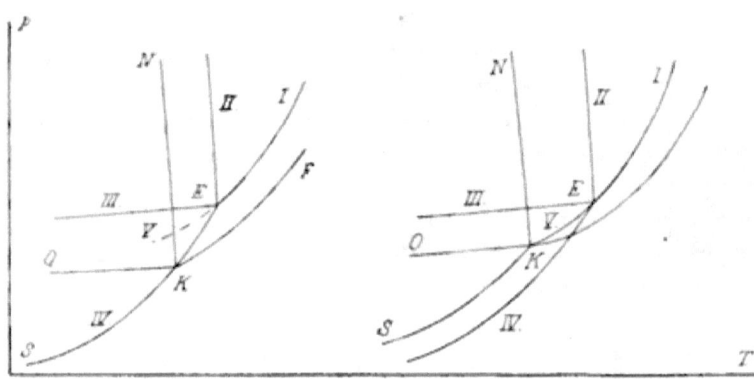

FIG. 45.

ruple point for the binary system is at E in both diagrams and the quintuple point for the ternary system at K. Along curves I, II, III and IV the solid phases are anhydrous sodium sulfate, solution and vapor ; hydrated and anhydrous sodium sulfate, and solution ; hydrated sodium sulfate, solution and vapor ; hydrated and anhydrous sodium sulfate, and vapor. Curve V represents the instable portion of curve I, the phases being anhydrous sodium sulfate, solution and vapor. The curves for the ternary system have the same lettering as in Fig. 44. In the left-hand diagram it is assumed that sodium sulfate effloresces under water when the pressure equals the dissociation pressure for pure hydrated sodium sulfate ; in the right-hand diagram it is assumed that the change takes place when the pressure equals that of the solution in equilibrium with anhydrous sodium sulfate. If the first hypothesis is the right one, EKS becomes a single curve and will always have the same values regardless of the nature of the third component. This seems to be the view adopted by Löwenherz[1]. An analogous case occured in the equilibrium between potassium chloride and water. There it was seen that if the partial pressure and the specific action of the potassium chloride were neglected, the pressure-temperature curves for ice, solution and vapor and for ice, salt and vapor were identical with the sublimation curve for ice.[2]

[1] Zeit. phys. Chem. **18**, 70 (1894).
[2] Cf. page 50.

Against this is to be set the fact, noticed with calcium chloride[1] that the dissociation pressure for a hydrated salt is a function of the dissociation products. A compound has a definite dissociation pressure only with respect to a given phase or set of phases just as a solution is saturated only with respect to a given phase or set of phases. This appears very clearly when the compound is a hydrated double salt which can effloresce in different ways. It is therefore probable, though not proved, that EK and KS are not parts of the same curve and that neither coincides with the dissociation curve of hydrated and anhydrous sodium sulfates. If the right-hand diagram in Fig. 45 represents the facts the curve KS will lie above curve IV or the presence of solid sodium chloride decreases the stability of hydrated sodium sulfate with respect to the anhydrous salt. Looked at in this way sodium chloride may be called a catalytic agent.

The solubility curve for a hydrated and anhydrous salt in a ternary system has been studied by Goldschmidt.[2] He makes the unnecessary and incorrect assumption that the solubility in water of an anhydrous salt is scarcely affected, if at all, by the addition of a non-electrolyte. From this premise he concludes that addition of a non-electrolyte increases the solubility of a hydrated salt. This is right as far as it goes; but it is inaccurately worded and is not a general statement of the relations. In a ternary system a single solid phase has not a definite solubility at a fixed temperature. It can exist in equilibrium with a series of solutions. The limiting solubility, or solubility in presence of another solid phase, can have but a single value for each temperature. Goldschmidt's proposition when accurately worded will read : If the salt concentration in the monovariant system, hydrated and anhydrous salt, solution and vapor, be equal to the solubility of the anhydrous salt in water, it will be greater than the solubility of the hydrated salt in water. This is only a special case of the well-known fact that one thing can not be equal to two different things. It is probable that the limiting solubility is determined in cases of this sort by pressure considerations and it is for this reason that Goldschmidt's results have been

---

[1] Cf. page 73.
[2] Zeit. phys. Chem. 17, 145 (1895).

referred to here rather than in the chapters on concentration and temperature. It is very important that some careful measurements be made to determine directly the pressure at which the efflorescence takes place. The chief objection to accepting the left-hand diagram in Fig. 45 is the behavior of hydrated double salts; but it is conceivable that these really come in a class by themselves and that one might lay down the rule that a solid phase affects the dissociation pressure of another solid phase only when it can react with one of the dissociation products.

Of special interest are the curves for the monovariant systems in which a hydrated double salt is one of the solid phases. As an example of a hydrated double salt which separates from solution only below a certain temperature, we may take the system, copper acetate, calcium acetate and water, in the neighborhood of $76°$. Roozeboom[1] has already pointed out the relative positions of the five curves at this quintuple point and his diagram is reproduced in Fig. 46.

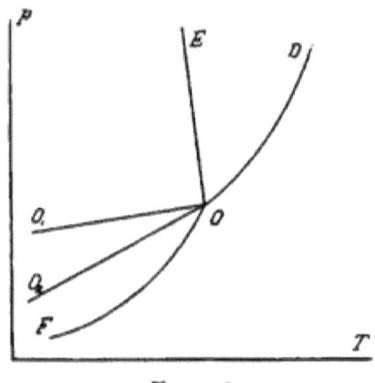

FIG. 46.

OD, OO_1, and OO_2 are the curves for two solid phases, solution and vapor. Along OD the solid phases are the two single acetates, $CuAc_2H_2O$ and $CaAc_2H_2O$; along OO_1, the hydrated double salt, $CuCaAc_4 8H_2O$, and crystallized copper acetate; along OO_2 the hydrated double salt and crystalized calcium acetate. The curve DO,

---

[1] Zeit. phys. Chem. **2**, 517 (1888).

if prolonged will lie below OO_2 for the reasons already given when discussing Fig. 44. OE is the curve for the double salt, the two single salts and solution. The curve slants to the left because the hydrated double salt occupies a larger volume than its dissociation products and must, therefore, be decomposed by pressure.[1] Spring and van 't Hoff[2] have found that, under an estimated pressure of six thousand atmospheres, the hydrated double salt is certainly converted into the single salts and solution at 40° and probably at 16°. The curve OF gives the simultaneous values for pressure and temperature when the double salt and the two single salts are in equilibrium with vapor. It is the dissociation curve for the hydrated double salt. This curve must lie above the dissociation curves for hydrated copper acetate or hydrated calcium acetate because these compounds can not begin to dissociate until the whole of the double salt has disappeared. It is to be noticed that OO_1 and OO_2 can be looked upon as dissociation curves for the hydrated double salt though not in the usual sense of the term. At pressures given by OO_1 the double salt effloresces under water with formation of copper acetate; at pressures given by OO_2 it effloresces under water with formation of calcium acetate while at pressures given by OF it effloresces with formation of the two single salts. Only this last is usually looked upon as a dissociation curve. At pressures between OF and OO_2 the hydrated double salt is stable in presence of solid calcium acetate; between OF and OO_1 in presence of solid copper acetate. It was seen in discussing Fig. 44 that KE might be looked upon as a continuation of SK; but such a view is untenable with a hydrated double salt because it can change under water in two ways. In this particular case it is doubly impossible because OO_1 OO_2 and OF all lie on the same side of the point O.

The temperature of the point O is a maximum temperature for this hydrated double salt whether in equilibrium with both solution and vapor or not. The conditions under which this occurs have been formulated by Roozeboom[3] as follows: " The quintuple point

---

[1] van 't Hoff and van Deventer, Zeit. phys. Chem. **1**, 173 (1887).
[2] Ibid. **1**, 227 (1887).
[3] Ibid. **2**, 517 (1887).

which one meets in the study of hydrated double salts is only a transition point (maximum temperature) for the double salt if the double salt contains more water of crystallization than the two components together and if the change into the components and solution is accompanied by contraction ; in all other cases the hydrated double salt can exist at higher and lower temperatures than the quintuple point."

As an example of a hydrated double salt which can exist in equilibrium with solution only above a certain temperature, Roozeboom has taken the double sulfate of sodium and magnesium.[1] In Fig. 47 is the pressure-temperature diagram for the equilibrium around the quintuple point at 21.5°. The monovariant systems containing both

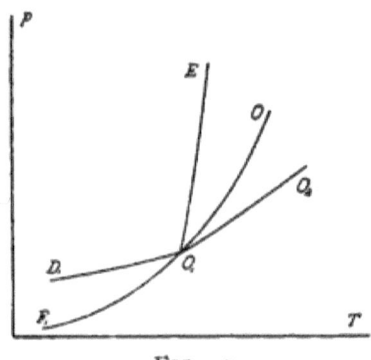

FIG. 47.

solution and vapor phases are represented by O_1O with the hydrated double salt, $Na_2Mg(SO_4)_2.4H_2O$, and hydrated sodium sulfate, $Na_2SO_4.10H_2O$, as solid phases ; by O_1O_2 with double salt and hydrated magnesium sulfate ; by O_1D with the two hydrated single salts as solid phases. These three curves end with the appearance of anhydrous sodium sulfate, magnesium sulfate with six of water and ice, respectively. Along O_1E_1 the phases are the two single salts, the double salt and solution. This curve slants to the right because the change of the single salts into double salt and solution is accompanied by an expansion of volume.[2] The curve O_1F_1 is the

---

[1] Zeit. phys. Chem. **2**, 513 (1888).
[2] van 't Hoff and van Deventer, Ibid. **1**, 173 (1887).

pressure-temperature curve for the three solid phases and vapor. It is clear from the diagram that the point O_1 is not a minimum temperature for the double sulfate of sodium and magnesium. A quintuple point may be a maximum or a minimum temperature for a given solid phase in equilibrium with solution and vapor, as has already been shown in the chapter on quintuple points ; but if it is not required that the solid phase shall coexist with solution and vapor, the quintuple point is not a minimum temperature for any solid phase, while it is a maximum temperature for a hydrated double salt only under the conditions laid down by Roozeboom, and the additional one—also stated explicitly by him—that the double salt is not stable at its melting point.

Having studied the arrangement of the curves round a quintuple point we are in a position to consider the pressure-temperature diagram, including all the monovariant systems in which a hydrated double salt is one of the solid phases. It has already been shown that the field for a ternary compound must be bounded by three curves, and may be by many more. The fields for the salts $K_2Mg(SO_4)_2.6H_2O$ and $CuCl_2.2KCl.2H_2O$ were each bounded by four curves ; there are no instances known as yet in which the field is bounded by three curves, but such a state of things is readily possible and must be taken into account because it is the simplest case. It may be well to mention that two salts which always crystallize in the anhydrous form from aqueous solution can not form a hydrated double salt unless this compound can exist in stable equilibrium with ice, solution and vapor. This can easily be seen by constructing the appropriate diagram.

In Fig. 48 are the pressure-temperature curves for a hydrated double salt which has no true melting point and which can exist in stable equilibrium with only three different solid phases, one of these being a hydrated salt. The figure is merely a combination of the last four diagrams. In order to make the relation between this and the concentration-temperature diagram more clear, the latter is shown with the same lettering in Fig. 49. If we call water A, one of the salts B and the other C, the hydrated salt will be AB and the hydrated double salt ABC.[1] In Fig. 48 the curve OD is the pressure-

---

[1] In the salts AB and ABC the letters denote the qualitative and not the quantitative relations.

temperature curve for B, C, solution and vapor; OO_1 for ABC, C, solution and vapor;[1] OO_2 for ABC, B, solution and vapor; O_1O_2 for ABC, $\overline{AB}$, solution and vapor; O_1D_1 for AB, C, solution and vapor; OE for ABC, B, C and solution; O_1E_1 for $\overline{ABC}$, AB, C and solution; OF for $\overline{ABC}$, B, C and vapor; O_1F_1 for ABC, AB, C and vapor. These curves have been considered under the systems, water with sodium and magnesium sulfates, and water with copper and calcium acetates. O_2 is also a quintuple point analogous to the one in Fig. 44. There are therefore the curves O_2D_2 for AB, B, solution and vapor; O_2E_2 for ABC, AB, B and solution; O_2F_2 for $\overline{ABC}$, AB, B and vapor. The relations of the three curves OF, O_2F_2 and O_1F_1

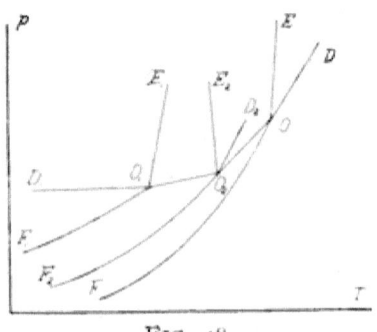

FIG. 48.

and the changes in the monovariant systems when the pressures fall below those of the curves are well worth a brief discussion. Whether any two of the three curves can intersect is not known. If they do not, O_1F_1 will always lie above O_2F_2 and this latter always above OF. Along OF the solid phases, as we have already seen, are ABC, B and C. If the pressure falls below the limiting value, the hydrated double salt will effloresce with formation of the salts B and C. Along O_2F_2 the solid phases are $\overline{ABC}$, $\overline{AB}$, and B. Decrease of pressure will cause the hydrated salt AB to effloresce with formation of the divariant system, ABC, B and vapor. This system will remain unchanged until the pressure reaches the value for OF at that tempera-

---

[1] This line has accidentally been omitted from the diagram.

ture, when the hydrated double salt will begin to change into B and
C. Along O_xF_1 the solid phases are $\overline{ABC}$, AB and C. Here the hydrated double salt will effloresce; but not with formation of B, as might be expected. It must be remembered that the salt AB contains more water of crystallization than the ternary compound ABC.[1]
When AB effloresces in presence of C, there is formed the hydrated double salt ABC. If there is an excess of $\overline{AB}$ relatively to C in the monovariant system, $\overline{ABC}$, $\overline{AB}$, C and vapor, a continued decrease of the external pressure will cause formation of the divariant system

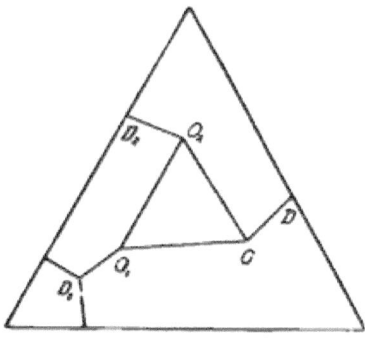

FIG. 49.

$\overline{ABC}$, $\overline{AB}$ and vapor; then of the monovariant system $\overline{ABC}$, $\overline{AB}$, B and vapor; the salt $\overline{AB}$ will next disappear; then the salt C will appear, and eventually $\overline{ABC}$ will disappear leaving only B, C and the vapor of A. If there is an excess of C relatively to $\overline{AB}$ in the monovariant system, $\overline{ABC}$, $\overline{AB}$, C and vapor, there will be formed the divariant system, $\overline{ABC}$, C and vapor, which will next change into the system $\overline{ABC}$, B, C and vapor, with final disappearance of $\overline{ABC}$. In other words, with excess of AB the system passes from O_xF_1 through O_xF, and OF; with excess of C the system passes from O_xF_1 through OF without intermediate formation of the monovariant system existing along O_xF_x. The salt $\overline{AB}$ effloresces in different ways and at different pressures, depending on its being in the presence of the salt B or the salt C.

[1] See page 195.

If the field for the ternary compound is bounded by four curves instead of three, this will make the difference in the pressure-temperature diagram that one of the lines OO_1, OO_2 and O_1O_2 will have a break and there will be a new quintuple point at this break. It must be distinctly remembered that the typical diagram given in Fig. 48 does not apply to the case where the hydrated double salt has a stable melting point. If any of the monovariant systems have a temperature maximum somewhere in the middle of the curve, representing them, it is clear that this will alter the diagram very considerably. Since no case is yet known of a ternary compound, made from two solids and a liquid which has a true melting point it is not worth while to consider the diagram for this special case.

The relations between pressure and temperature have been determined quantitatively for only one system. Vriens'[1] has studied the equilibrium between cupric chloride, potassium chloride and water from this point of view. One of the interesting things about his results is that the curve for $CuCl_2 2KCl 2H_2O$, $CuCl_2 2H_2O$, $CuCl,KCl$ and vapor lies above the curve for $CuCl_2 2KCl 2H_2O$, KCl, $CuCl,KCl$ and vapor while this latter probably lies above the undetermined curve for $CuCl_2 2H_2O$, $CuCl$, $CuCl,KCl$ and vapor. This seems like a contradiction of the theoretical results already obtained; but the difference is due to the formation of an anhydrous double salt $CuCl,KCl$. Starting from the system $CuCl_2 2KCl 2H_2O$, $CuCl_2 2H_2O$, $CuCl,KCl$ and vapor, and decreasing the external pressure, there will be disappearance of copper dipotassium chloride and hydrated copper chloride with formation of copper potassium chloride. If the hydrated double salt be present in excess, hydrated copper chloride will be the first phase to disappear, forming the divariant system, $CuCl_2 2KCl 2H_2O$, $CuCl,KCl$ and vapor. At a yet lower pressure the hydrated double salt will effloresce with formation of potassium chloride and copper potassium chloride, the pressure remaining constant so long as the monovariant system, $CuCl_2 2KCl 2H_2O$, KCl, $CuCl,KCl$ and vapor is present. If hydrated copper chloride were originally in excess, instead of the hydrated double salt, this latter would be the first to disappear, leaving the divariant

---

[1] Zeit. phys. Chem. **7**, 194 (1891).

system, CuCl$_2$2H$_2$O, CuCl,KCl and vapor. This will remain in stable equilibrium until the pressure falls below the value for the system CuCl$_2$2H$_2$O, CuCl,KCl and vapor. If we start with the hydrated double salt, alone or in presence of either or both of the salts, KCl and CuCl,KCl, the hydrated double salt will effloresce with formation of potassium chloride and copper potassium chloride. It will be noticed that the reactions, and the pressures at which they take place, are functions of the nature and relative amounts of the solid phases originally present. In this particular system there is always disappearance of the hydrated double salt. This is not necessarily the case. While there are no experimental data as yet, it seems fairly certain that a mixture of magnesium sulfate heptahydrate and sodium sulfate decahydrate will effloresce with formation of the hydrated double salt, Na$_2$Mg(SO$_4$)$_2$4H$_2$O. If the sodium salt is present in excess, this will then effloresce forming the anhydrous salt, and not till this change is completed will the hydrated double salt begin to dissociate. If there is an excess of magnesium sulfate heptahydrate, this salt will effloresce with formation of the hexahydrate and the latter will probably effloresce before the double salt does. In spite of the diametrically opposite behavior of these two hydrated double salts, the same rule applies to both cases and to all others where only one of the components is measurably volatile. Two solid phases containing three components will effloresce with formation of that solid phase which can exist in equilibrium with them at the next higher quintuple point.

When an anhydrous and a hydrated salt combine to form a hydrated double salt with addition or subtraction of water, there seems at first no reason why there should not be a quintuple point at which these three solid phases could be in equilibrium with solution and vapor, yet this is not possible. To take a concrete case let us assume that lead and potassium iodides form no anhydrous double salt, and only one hydrated double salt, PbI,KI2H$_2$O. If the nonvariant system, hydrated lead iodide, potassium iodide, lead potassium iodide, solution and vapor, can exist it will be possible to have these three salts in equilibrium with vapor over a series of temperatures. A moment's consideration will show that there is no way in which this system can effloresce without forming a new solid phase and thus a nonvariant system capable of existing over an indefinite range of temperature. Since this is impossible, it follows that a quintuple point with these three salts as solid phases is impossible and that another solid phase must appear before this point is reached. Under the conditions assumed to exist, the new phase would be lead iodide either anhydrous or with one of water. As a matter of fact, it is probable that lead and potassium iodides form a second hydrated double salt and it is this phase which appears.[1] The hydrated double chloride of copper and potassium can be made from potassium chloride and hydrated cupric chloride without addition or subtraction of water. Here it is known that the anhydrous double salt $CuCl_2KCl$ appears as solid phase and that copper dipotassium chloride, potassium chloride and hydrated cupric chloride cannot co-exist in equilibrium with solution and vapor.

The efflorescence of copper dipotassium chloride has already been discussed; but the effect of diminished pressure on the double iodide of lead and potassium needs some consideration. If the solid phases at one of the quintuple points are lead potassium iodide, hydrated and anhydrous lead iodide and, at the other, lead potassium iodide, potassium iodide and anhydrous lead iodide, a mixture of hydrated double salt and hydrated lead iodide will effloresce to hydrated double salt, hydrated and anhydrous lead iodide. When the hydrated lead iodide has entirely disappeared the hydrated double salt will begin to effloresce, forming anhydrous lead iodide and potassium iodide. The pure hydrated double salt or a mixture of this with potassium iodide will effloresce with the formation of the monovariant system, hydrated double salt, anhydrous lead iodide and potassium iodide, the double salt finally disappearing entirely.

If the solid phases at one of the quintuple points are lead potassium iodide, a second double salt with or without water of crystallization and hydrated lead iodide and, at the other, the two double salts and potassium iodide, a mixture of lead potassium iodide and hydrated lead iodide will effloresce with formation of the second double salt at the expense of the first, assuming that the ratio of lead to potassium is the same in the two double salts. When lead

---

[1] Schreinemakers, Zeit. phys. Chem. **10**, 471, (1892).

potassium iodide has entirely disappeared, the hydrated lead iodide will begin to effloresce. If the second double salt contain water of crystallization and an anhydrous lead potassium iodide cannot exist, the second double salt will begin to effloresce when all the hydrated lead iodide is gone, and there will be formed the divariant system, anhydrous lead iodide, potassium iodide and water vapor. A mixture of lead potassium iodide, PbI,KI2H,O, and potassium iodide will effloresce, on the same assumptions, with formation of the second double salt at the expense of the first, the second efflorescing in its turn, and the final result being the divariant system, anhydrous lead iodide, potassium iodide and water vapor. The first double salt, if pure, will form the second salt[1] and then this will change into the two single iodides. If the first double salt contain less potassium iodide than the second, it will change to the latter and lead iodide. It is evident that an examination of the products of efflorescence will give definite information on the nature of the phases existing at quintuple points which can not be easily investigated in the usual manner.

It must be remembered that all this reasoning is based on the assumption that the dissociation curves do not intersect. Since two adjacent dissociation curves always have two solid phases in common, an intersection would form a new quintuple point at which four solid phases would be in equilibrium with vapor. If the four solid phases be denoted by the letters w, x, y and z respectively, the five curves meeting in the quintuple point will represent the simultaneous pressures and temperatures for the five monovariant systems; w, x, y and vapor; x, y, z and vapor; y, z, w and vapor; z, w, x and vapor; w, x, y and z. No instance of such a quintuple point has yet been observed. If such an one shall ever be found, there will then be a temperature below which a given compound can not exist. Such cases occur in one component systems, where one solid modification changes into another with evolution of heat and concentration of volume. In binary systems no such case has yet been found, so that it is not surprising that none is known for ternary systems.

---

[1] Assuming that the ratio of lead to potassium is the same in the two double salts.

Before leaving the subject of pressures, reference should be made to one other pressure-temperature curve for three solid phases and vapor. By heating together lead oxide and ammonium chloride, there are formed, in addition to the two original substances, lead oxide hydrochloride, PbOHCl, and ammonia gas. The components are three in number: lead oxide, ammonia and hydrochloric acid. It was found by Isambert[1] that the three solid phases and vapor can exist over a range of temperatures but under only one pressure for each temperature.

[1] Comptes rendus, **102**, 1313 (1886).

CHAPTER XIV

SOLID SOLUTIONS

No quintuple points have been determined for systems in which solid solutions are possible. This is not because such points are rare, but because no one has been interested in them. While all systems containing two salts and water can form one or more nonvariant systems under suitable conditions of temperature and pressure when no solid solutions are possible, this is not necessarily true if this restriction be dropped. If two substances form a single continuous series of solid solutions, it has already been seen that a quadruple point is impossible. Such a system plus water could form a quintuple point only in case the water decomposed the solid solutions.

Though no ternary systems containing solid solutions have been studied in detail, there are some rather haphazard measurements by Le Chatelier[1] upon mixtures of melted salts, while isotherms have been determined by Roozeboom,[2] Stortenbeker,[3] and others.[4] More interesting from the quantitative than the qualitative point of view are the investigations of Küster.[5] Ether is soluble to a certain extent in water and in rubber, while the latter substances are practically non-miscible. Küster studied the divariant system, vapor, solution of ether in rubber and solution of ether in water. The measurements extend only over a limited range of concentrations, and the amount of water which was carried into the solid phase by the ether was not determined, though it is safe to assume that the amount was less than would be calculated from the solubility of water in ether. Similar measurements have been made with starch, iodine and water,

---

[1] Comptes rendus, **118**, 415 (1894).
[2] Zeit. phys. Chem. **8**, 531 (1891); **10**, 147 (1892).
[3] Ibid. **16**, 250; **17**, 643 (1895).
[4] Pock, Ibid. **12**, 661 (1893); Muthmann and Kuntze, Zeit. **Kryst. 23,** 368 (1894).
[5] Zeit. phys. Chem. **13**, 445 (1894).

showing that the iodine forms a solid solution with starch and that the system is therefore divariant.[1] Walker and Appleyard[2] have shown that picric acid forms a solid solution with silk, and they have studied the distribution of picric acid between silk and water. Curiously enough, after giving a most satisfactory proof that a solid solution is formed, they draw the conclusion that this is not the case. In most other experiments with dye-stuffs there are more than three components, because the bath is acidified in order to get better results. Other cases of divariant systems in which one of the phases is a solid solution have been studied by Schmidt[3] and by van Bemmelen.[4] Charcoal possesses the power of taking coloring matter out of solutions and it was shown by Graham[5] that metallic salts are also absorbed. In both these cases there can be little doubt but that solid solutions are formed though adsorption phenomena may take place to a minor extent. Curiously enough, almost no work has been done in this field of late years and we are very ignorant of the laws describing these phenomena. We do not even know whether the behavior of charcoal and filter paper is analogous to that of glass wool[6] though this is probably not the case.

[1] Liebig's Annalen, **283**, 360 (1894).
[2] Jour. Chem. Soc. **69**, 1334 (1896).
[3] Zeit. phys. Chem. **15**, 56 (1894).
[4] Jour. prakt. Chem. **23**, 324, 379 (1881); Zeit. phys. Chem. **18**, 331 (1895).
[5] Pogg. Ann. **19**, 139 (1830).
[6] Cf. Ostwald, Lehrbuch I, 1084.

CHAPTER XV

ISOTHERMS

Having already considered the changes of concentration with the temperature for a series of typical systems, we can now take up the more special cases of the changes of concentration when one of the salt components is added continuously to a solution kept at a constant temperature. If we ignore the water of crystallization, which the solid phases may contain, there are eight cases to consider[1]:

Case I. The two salts form neither double salts nor mix crystals. Examples of this are silver nitrate and acetate, sodium chloride and nitrate.

Case II. The salts form a continuous series of solid solutions. An example of this is to be found in ammonium and potassium sulfates, permanganate and perchlorate of potassium.

Case III. There are formed two series of solid solutions. Examples: Thallium and potassium chlorates, cobalt and copper sulfates, iron and magnesium sulfates.

Case IV. The solid phases are one double salt and two single salts. Examples: Potassium and copper sulfates, sodium and magnesium sulfates.

Case V. The solid phases are two double salts and two single salts. Examples: Copper and potassium chlorides.

Case VI. The solid phases are one salt, one double salt and one series of solid solutions. Example: Ammonium and ferric chlorides.

Case VII. The solid phases are one double salt and two series of solid solutions. Examples: Potassium and silver nitrates, ammonium and silver nitrates, potassium and silver chlorates, potassium and sodium sulfates.

---

[1] Cf. Roozeboom, Zeit. phys. Chem. **8**, 519 (1891); **10**, 158 (1892); Schreinemakers, Ibid. **11**, 88 (1893); Stortenbeker, Ibid. **17**, 643 (1895). For the earlier literature on the subject, see Ostwald, Lehrbuch I, 1072–1079.

Case VIII. There are formed three series of solid solutions. Examples : Magnesium and copper sulfates, zinc and copper sulfates.

The typical diagrams for these eight cases are given in Fig. 50. The ordinates are reacting weights of one salt and the abscissae reacting weights of the other salt, both in a constant amount of water.[1]

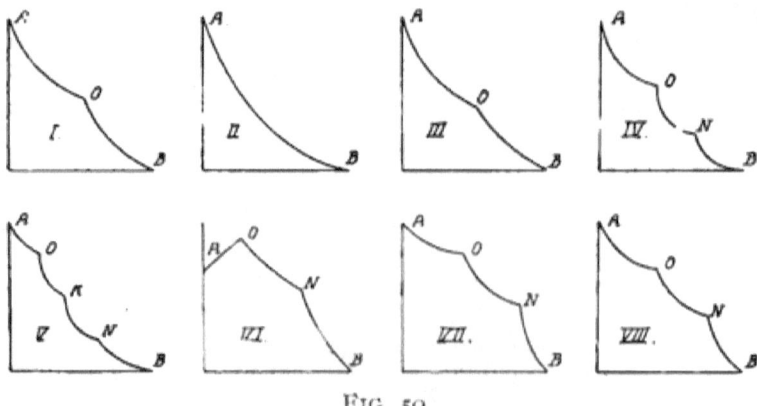

FIG. 50.

In most cases addition of a salt to a saturated solution of a second salt, either the acid or the basic radicle being common, produces a decrease in the solubility of the second salt. All the diagrams except the one for Case VI are drawn to show this; but it should be kept in mind that this is not necessarily true. Exceptions are to be found with potassium and sodium nitrates,[2] potassium and lead nitrates.[3] This affects the direction of the curves but not the movement along them.

In the diagram for Case I, the salt A is solid phase along AO and the salt B along OB. Addition of B produces a precipitation of A, and the system passes along the curve AO, becoming poorer in A and richer in B until, at the point O, the solution is saturated with respect to B. There is then present the monovariant system, two salts, solution and vapor. Further addition of B produces no change in the concentration because the solution is already saturated with

---

[1] Cf. Schreinemakers, Zeit. phys. Chem. **9**, 67 (1892).
[2] Nicol. Phil. Mag. (5) **31**, 369 (1891).
[3] Le Blanc and Noyes, Zeit. phys. Chem. **6**, 385 (1890).

respect to B, and therefore the added amount merely increases the quantity of that component as solid phase. In like manner continued addition of A to a solution saturated with respect to B produces a precipitation of B, the system passing along BO. At O further addition of A has no effect because the solution is saturated with respect to A. When the two salts do not form double salts or solid solutions, neither can displace the other completely, and the equilibrium reached by adding B to a solution of A is the same as that reached by adding A to a solution of B. Quantitative determinations of several isotherms of this class have been made by Bodländer[1] and by Nicol.[2]

In the diagram for Case II the solid phase along AB is a solid solution with continuously varying concentration. Addition of B produces a change in the composition of the two solutions, solid and liquid, the ratio of B to A increasing in both as the system passes along the line AB. The final result is a pure solution of B to within any desired degree of accuracy, though an infinite quantity of B must be added to attain this. Addition of A to a solution saturated with respect to B produces the reverse change, practically the whole of B being precipitated. When the two salts form a continuous series of solid solutions, either can precipitate the other practically completely from the solution. In practice this result is reached more quickly if the crystals are removed as fast as formed. If they are left at the bottom of the solution, one of the salts will have to be added until the total quantity of the first is infinitely small in comparison with the amount of the second. An isotherm for ammonium and potassium sulfates, a pair of salts belonging to this class, has been determined by Fock[3] while Muthmann and Kuntze have studied the permanganate and perchlorate of potassium.[4]

In the diagram for Case III the solid phase along AO is a solid solution containing the salt A in excess, and along BO a solid solution containing an excess of the salt B. Addition of B to a solution

---

[1] Zeit. phys. Chem. **7**, 360 (1891).
[2] Phil. Mag. (5) **31**, 369 (1891).
[3] Zeit. phys. Chem. **12**, 661 (1893).
[4] Zeit. Kryst. **23**, 368 (1894).

saturated with respect to A produces a change in the solutions, solid and liquid, until the concentration denoted by the point O is reached. The solution is then saturated with respect to both sets of solid solutions. Further addition of B causes a decrease in the amount of the solid phase in which A is solvent and an increase in the quantity of the other set of crystals, the concentrations of the three solutions, two solid and one liquid, remaining unchanged. When the crystals containing an excess of A are entirely consumed, addition of B causes a change in the concentrations of both the remaining solutions, the systems passing along OB. The final result is, to all intents and purposes, a solution saturated with respect to B only. Exactly the opposite changes take place on adding A to a saturated solution of B. The system moves along the curve BO, stays at O until the crystals containing an excess of B have disappeared and then passes along AO, ending with what is a pure solution of A from the standpoint of quantitative analysis. When the salts form two sets of solid solutions either salt can displace the other completely. An isotherm for potassium and thallium chlorates has been determined by Bakhuis Roozeboom[1] and for other pairs of salts by Stortenbeker[2]. For a lecture experiment, the sulfates of cobalt and copper are excellent because the two sets of mix crystals have different colors.

In the diagram for Case IV the salt A is solid phase along AO and the salt B along BN; while, along ON, there is the double salt. Addition of B to a saturated solution of A causes the system to pass along the line AO. At O the double salt appears as solid phase and further addition of B causes no change in the concentration until all of A in the solid phase has been converted into double salt. When this has happened the system will pass along the line ON until, at N, B appears as solid phase and further addition of this component produces no change in the concentration. Addition of A to a solution

---

[1] Zeit. phys. Chem. **8**, 530 (1891). The compositions of the two solid solutions at the point O are 36.3 and 97.93 reacting weights of potassium chlorate in one hundred reacting weights of the solid solutions, the temperature being 10°.

[2] Ibid. **16**, 250, 1895. It is not yet certain, that there are not two classes under this one heading, two series of solid solutions and limited solubilities, two series of solid solutions with unlimited solubilities.

saturated in respect to B causes the system to pass along BN and then along NO, the concentration remaining constant at N until the crystals of B have been entirely changed into double salt. At O the salt A appears as solid phase and no further change of concentration is possible. When the solid phases along the isotherm are the two single salts and a double salt, neither salt can precipitate the other completely from the solution and the equilibrium reached by adding B to a solution of A is not the same as that reached by adding A to a solution of B. It is immaterial whether the double salt is decomposed by water or not. Sodium and magnesium sulfates furnish an excellent example of this case, the double salt being decomposed by water between 21.5° and 25° and not decomposed above the latter temperature.[1] Some points on the curve for potassium and copper sulfate have been determined by Trevor.[2] This is a case where the double salt is not decomposed by water at any temperature so far as is now known. As an illustration of a double salt decomposed by water at all temperatures we have the double iodide of lead and potassium studied by Schreinemakers.[3] Though these three double salts behave so differently towards water, the three systems show the same changes when either salt component is added continuously.

In the diagram for Case V the salt A is solid phase along AO and the salt B along BN, while one of the double salts exists along OK and the other along NK. Addition of B to a solution of A causes the system to pass from A to N, the concentration remaining constant at O until all the crystals have been converted into the first double salt, and constant again at K until the crystals of the first double salt have been changed into those of the second double salt. When the system has reached the point N, further addition of B produces no change beyond an increase in the amount of B present as solid phase. When the solid phases along the isotherm are the two single salts and the two double salts, neither salt can displace the other completely, and the equilibrium reached by adding B to a solution of A is not the same as that reached by adding A to a solution

---

[1] Roozeboom, Zeit. phys. Chem. **2**. 513 (1888).
[2] Ibid. **7**. 460 (1891).
[3] Ibid. **10**. 467 (1893).

of B. An example of this form of isotherm is furnished by copper and potassium chloride between 56° and 92°.[1]

Only one instance of Case VI has yet been observed, the salts being ammonium and ferric chloride.[2] Along AO the solid phase is $FeCl_3 6H_2O$; along ON the double salt $2NH_4ClFeCl_3H_2O$, and along NB a series of solid solutions. Addition of ammonium chloride to a saturated solution of ferric chloride produces an increase in the solubility of the latter and the system passes along the curve AO. At O the solution becomes saturated with respect to the double salt and further addition of ammonium chloride brings about no change in the concentration until all the crystals of ferric chloride have been converted into crystals of the double salt. The system then passes along the curve ON until, at N, there is formed the monovariant system, double salt, the first term of the series of solid solutions, solution and vapor. On adding ammonium chloride, the crystals of the double salt disappear while the amount of the solid solution increases, the concentration of the liquid and the solid solutions remaining constant. When the double salt has completely disappeared the concentrations of the two solutions change with addition of ammonium chloride, approaching as a limit a pure saturated solution of ammonium chloride. The double salt is decomposed by water. The solid solutions contain all three components. Addition of ferric chloride to a solution saturated with respect to ammonium chloride produces formation of solid solutions and then disappearance of solid solutions with formation of double salt, the system passing along the curves BNO. When the point O is reached further addition of ferric chloride has no effect upon the concentration of the solution. When the order of solid phases along the isotherm is single salt, double salt and a solid solution, the second salt can precipitate the first completely but the first can not do this to the second.

In the diagram for Case VII the solid phase along AO is a solid solution with A as solvent, along BN a solid solution with B as solvent while along ON double salt crystallizes. Addition of B to a solution of A causes a continuous change in the concentrations of

[1] Meyerhoffer, Zeit. phys. Chem. **5**, 97 (1890).
[2] Roozeboom, Ibid. **10**, 147 (1892).

the solid and liquid phases until the system reaches O and the double salt crystallizes. The concentration remains constant while the last term of the series of solid solutions is being converted into double salt. When this is finished the system passes along ON and NB, the concentration remaining constant at N while the double salt changes into the new series of solid solutions. The system finally approaches a saturated solution of pure B as a limit. Adding A to a solution of pure B, the order of events is reversed and the system approaches a pure, saturated solution of A as a limit. When the order of solid phases is solid solution, double salt and solid solution, either salt can precipitate the other completely. No isotherm of this class has been studied because, in the substances discovered so far, the range of the two solid solutions is so limited that it requires a microscopic examination to prove that it is not the pure single salt which separates. Retgers has shown that the nitrates of potassium and silver, the nitrates of ammonium and silver,[1] the chlorates of potassium and silver,[2] and the sulfates of sodium and potassium[2] belong under this heading.

In the diagram for Case VIII, solid solutions occur along AO, ON and NB. Adding B to the saturated solution of A causes the system to pass along the curve AONB, while an addition of A to a solution saturated with respect to B causes the system to traverse the path BNOA. In both cases the final result is a pure solution of the component added, to within any desired degree of accuracy; and in both cases there is a period of constant concentration at O and at N, while one solid phase is being replaced by the other. Examples of this are magnesium and copper sulfates, zinc and copper sulfates. Retgers[3] has determined the limiting concentrations of the crystals at the points O and N, but there are no determinations of the concentrations of the liquid solutions at these points. This will soon be remedied as Stortenbeker[4] has already announced a paper on this subject. Stortenbeker[5] points out that three series of solid solutions can occur in at least two ways. There can be three different kinds

---

[1] Zeit. phys. Chem. **4**, 611 (1889).
[2] Ibid. **5**, 446 (1890); Ibid. **6**, 226 (1890).
[3] Ibid. **15**, 573 (1894).
[4] Ibid. **17**, 650 (1895). [5] Ibid. **17**, 646 (1895).

of crystals, as in the two instances just referred to. The same result can be reached with only two kinds of crystals when a continuous series of solid solutions is cut into two parts by a second series. An example of this is probably to be found in the sulfates of copper and manganese. Under these circumstances AO and NB would be parts of the same continuous curve. This case, as well as IIIb, are derived from Case II by addition of another series of solid solutions. In IIIb the new series cuts off one end of the other curve while here it is the middle of the curve which disappears. When there are three series of solid solutions, either salt can precipitate the other completely. It is of no importance, so far as this point is concerned, whether the three series are all different or whether the two end ones have the same crystalline form.

In the eight cases which have been considered there have been instances where neither salt could precipitate the other, where only one could and where either could. In the instance where neither salt could precipitate, there are examples where the same equilibrium is reached no matter which salt is added continuously, and examples where a different equilibrium is reached. This may seem like a state of confusion; but Roozeboom[1] has pointed out that one simple rule covers all cases. On adding a component continuously a final equilibrium will be reached only when the solution is saturated with respect to that component. A corollary of this is that one salt can precipitate the other completely only when the second dissolves in the first to form a solid solution.

[1] Zeit. phys. Chem. **10**, 161 (1892).

CHAPTER XVI

FRACTIONAL EVAPORATION

The change in the system when one of the salt components is added continuously at constant temperature is independent of the absolute or relative solubility of either of the salts, but this is no longer the case when water is added to or removed from the solution. It will therefore be necessary to consider by itself the changes in the nature of the solid phase when water is withdrawn from the solution at constant temperature, the crystals being removed as fast as formed.

Under systems composed of two liquid components we have already considered the subject of fractional distillation. Here, the problem is the somewhat similar one of fractional evaporation of the mother liquor at constant temperature.[1] The relation between the liquid and the solid phase will be seen most clearly if a rectangular diagram be used, in which the ordinates represent reacting weights of the salt A in one hundred reacting weights of the salts A and B in the solution while the abscissae denote reacting weights of the same salt in one hundred reacting weights of the two salts in the solid phase.[2] The lower left hand corner of the diagram represents a saturated solution of B; the upper right hand corner a saturated solution of A. The horizontal double lines show that the two solid phases which are connected can coexist in equilibrium with the solution. These lines are not necessary to the diagram and could be omitted. The dotted diagonal shows the solutions in which the relative proportions of the two salts are the same as in the solid phase. It is clear that if the relative concentration of A is greater in the liquid than in the solid phase this difference will increase as more of the solid phase is formed and the solution will become relatively richer

---

[1] Cf. Meyerhoffer, Die Phasenregel, 49; **Schreinemakers,** Zeit. phys. **Chem. 11,** 81 (1893).

[2] Cf. Roozeboom, Zeit. phys. Chem. **8,** 521 (1891); Stortenbeker, Ibid. **16,** 257 (1895).

with respect to A. If, at any moment, the point representing the state of the system lie above the dotted diagonal, the solution will concentrate on evaporation towards a pure solution of A; if it lie below the line, it will concentrate towards pure B. In Fig. 51 are given most of the possible diagrams for fractional evaporation at constant temperature of a solution containing two salts. The Roman numerals refer to the cases which the pairs of salts illustrate.

Case I. The solid phases are the two single salts either hydrated or not hydrated. The diagram shows that where B is solid phase all the points lie above the diagonal and the solution concentrates toward A. All the solutions in equilibrium with A as solid phase lie below the diagonal, and these solutions therefore concentrate towards B. The final result of fractional evaporation in this case will be simultaneous separation of A and B, and when this begins to take place the solution will go dry without change of concentration. As an example of this take $A = NaCl$, $B = NaNO_3$, or *vice-versa*.

Case II. The salts form a continuous series of solid solutions. There are three possibilities:

a. The concentration of A in the solution may always be greater than the concentration of A in the solid phase. The solution changes finally to a solution containing only A. This happens when $A = (NH_4)_2SO_4$, $B = K_2SO_4$.[1]

b. The concentration of A in the solution may be greater and then less than the concentration of A in the crystals. The solution will pass to the point x where the two concentrations are equal and will then go to dryness without change. This happens when $A = KMnO_4$, $B = KClO_4$, or *vice-versa*.[2]

c. The concentration of A in the solution may be less and then greater than the concentration in the crystals. Solutions to the left of the point x would concentrate to pure B; solutions to the right to pure A. No instance of this has yet been studied.

Case III. These are two series of solid solutions. There are three possibilities:

[1] Fock, Zeit. phys. Chem. **12**, 661 (1893).
[2] Muthmann and Kuntze, Zeit. Kryst. **23**, 368 (1894).

Three Components

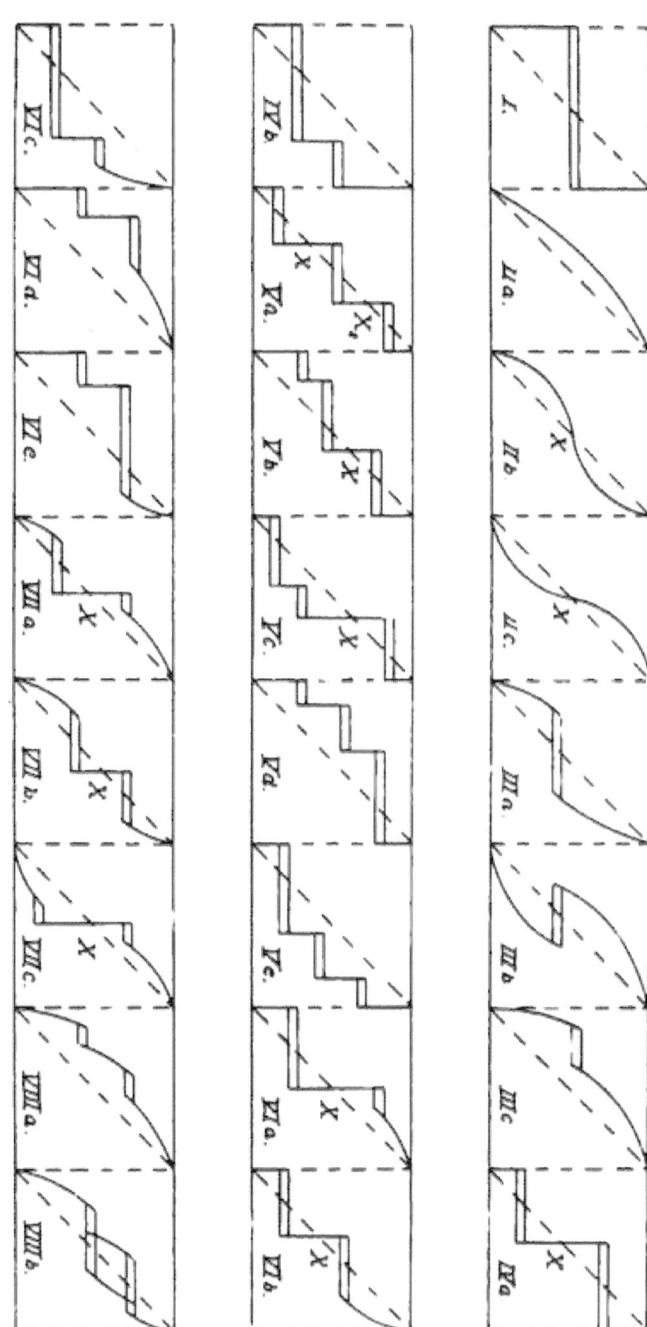

Fig. 51.

a. The concentration of A in the solution may be greater and then less than the concentration in the crystals. The solution will pass to the concentration at which the two kinds of crystals precipitate simultaneously, and will evaporate to dryness without further change. This happens when A = KClO$_3$, B = TlClO$_3$, or *vice-versa* ;[1] also with the stable portions of the system containing magnesium and ferrous sulfates or manganese and cobalt chlorides.[2]

b. The concentration of A in the solution may be less and then greater than the concentration in the crystals. Solutions corresponding to the first part of the curve would concentrate to pure B, those corresponding to the second part of the curve to pure A. The labile modifications of the system containing ferrous and magnesium sulfates illustrate this form of curve.[2]

c. The concentration of A is always greater in the solution than in the crystals. The solution finally changes to pure A. This happens when A = CoSO$_4$,7H$_2$O, B = CuSO$_4$,5H$_2$O ;[3] also when A = CuCl$_2$, 2KCl2H$_2$O, B = CuCl$_2$,2NH$_4$Cl2H$_2$O.[4]

Case IV. The solid phases are two single salts and a double salt. There are two possibilities :

a. The double salt is not decomposed by water. The concentration of A in the solution is greater, then less, then greater and finally less than its concentration in the crystals. Solutions containing less of A than the double salt will pass to the solution from which the double salt and the salt B crystallize simultaneously. Solutions containing a larger proportion of A than the double salt will change to the solution from which the double salt and salt A crystallize simultaneously while a pure solution of the double salt evaporates to dryness without change of concentration. This happens with potassium and copper sulfates ;[5] also with sodium and magnesium sulfates between 25° and 30°.[6]

b. The double salt is decomposed by water. The concentration

---

[1] Roozeboom, Zeit. phys. Chem. **8,** 532 (1891).
[2] Stortenbeker, Ibid. **16,** 250 (1895).
[3] Unpublished determination by K. K. Bosse.
[4] Fock, Zeit. phys. Chem. **12,** 658 (1893).
[5] Trevor, Ibid. **7,** 468 (1891).
[6] Roozeboom, Ibid. **2,** 518, (1888).

of A in the solution is greater and then less than the concentration of A in the solid phase. All solutions come finally to the solution from which the double salt and salt B crystallize simultaneously. This occurs when $A = Na_2SO_4 10H_2O$, $B = MgSO_4 7H_2O$ at about $24°$;[1] $A = KCl$, $B = CuCl_2 2H_2O$ at about $40°$;[2] $A = CuCl_2 2H_2O$, $B = LiCl$; $A = PbI_2 2H_2O$, $B = KI$ at about $69°$.[3] The behavior of the system containing lead and potassium iodides is so curious that it is worth considering for a moment in detail. Meyerhoffer[4] pointed out that, at this temperature, the ratio of the two salts happened to be the same for the two monovariant systems, with lead iodide and double salt and with potassium iodide and double salt as solid phases. Schreinemakers[5] took this up and showed that it was necessary to know the form of the isotherm in order to predict the changes and that in this particular case the isotherm was not a line of constant relative concentration for the two salts as Meyerhoffer had assumed. The diagram for this isotherm is given in Fig. 52 where the concentration of lead iodide in the solid and liquid phase is given. The line for the solutions in equilibrium with the double salt is really double, there being two sets of these solutions in which the ratio of

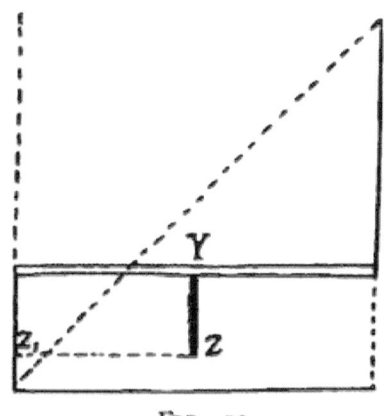

FIG. 52.

---

[1] Meyerhoffer, Zeit. phys. Chem. **5**, 103, 1890.
[2] Meyerhoffer, Sitzungsber. Akad. Wiss. Wien, **101** IIb, 599, 1892.
[3] Schreinemakers, Zeit. phys. Chem. **9**, 65, 1892.
[4] Ibid. **9**, 645, 1892; Die Phasenregel, 52.
[5] Zeit. phys. Chem. **10**, 475, 1892.

the two salts is constant and only the quantity of water varies. Starting from the monovariant system, lead iodide, double salt, solution and vapor, if we evaporate at the constant temperature of about 69° we shall have the lead iodide converted into double salt, the concentration of the solution remaining unchanged. When the lead iodide has entirely disappeared the system will pass along the line YZ with precipitation of double salt. On reaching Z the system will pass back along the other portion of the line ZY and the double salt will redissolve until the amount of lead iodide in one hundred of the two salts is again expressed by the point Y. The ratio of the water to the two salts is much less than before and the solution is saturated both with respect to potassium iodide and the double salt. The solution will now evaporate to dryness without change of concentration. When the crystals are not removed, the order of changes is: precipitation of lead iodide, conversion of lead iodide into double salt, precipitation of the double salt, solution of double salt and, finally, simultaneous precipitation of double salt and potassium iodide. This has all been pointed out very clearly by Schreinemakers in the paper referred to; but the same changes will not take place with fractional evaporation at constant temperature, the crystals being removed as fast as formed. Under these circumstances, there will first be precipitation of lead iodide and then of double salt as before until the system reaches the concentration represented by the point Z. If the crystals of the double salt are removed, as by hypothesis, the system cannot pass along the line ZY because there is no double salt present to dissolve and the ratio of lead iodide to potassium iodide cannot be increased by precipitation of the latter salt because the solution is not saturated with respect to it. The only thing possible is that the solution becomes an unsaturated one with respect to everything and remains so until at Z_1 potassium iodide separates and the solution finally concentrates to the point where double salt and potassium iodide crystallize simultaneously[1]. The order of changes when one starts from a solution rich in lead iodide and performs a fractional evaporation at the constant temperature of

---

[1] While there seems to be no flaw in the deduction, the conclusion is so unexpected that it should certainly be tested experimentally.

69° will be precipitation of lead iodide, precipitation of double salt, formation of an unsaturated solution, precipitation of potassium iodide and simultaneous precipitation of lead iodide and double salt. If the line YZ, had projected upwards instead of downwards it seems probable that the order would have been, precipitation of lead iodide, formation of unsaturated solution and then simultaneous precipitation of potassium iodide and double salt.

Returning from this digression we can now take up Case V with its five subheads, the solid phases being the two single salts and two double salts.

a. Neither of the double salts is decomposed by water. Solutions containing less of A than the first double salt concentrate to the solution from which the first double salt and salt B crystallize simultaneously; solutions containing more of A than the first double salt and less than the second, concentrate to the solution from which the two double salts separate simultaneously; solutions containing more of A than either double salt, concentrate to the second double salt and salt A, while the two solutions in which the salts are present in the same ratio as in the two double salts, evaporate to dryness without change of concentration. An example of this is to be found with cupric and tetrethyl ammonium chlorides at 31°.[1]

b. One of the double salts is decomposed by water, the second is not and the concentration of A is less in the first double salt than in the solutions in equilibrium with it. Solutions containing less of A than the second double salt concentrate to the solution in equilibrium with the two double salts; solutions containing relatively more of A than the second double salt concentrate to the solution from which the second double salt and the salt A separate, while the solution containing the two salts in the same proportion as the second double salt does not change concentration with loss of water at constant temperature. This system is realized at 80° when $A = CuCl_2, 2H_2O$ and $B = KCl$.[2]

c. One of the double salts is decomposed by water, the second is not; and the concentration of A in the first or decomposable double

---

[1] Meyerhoffer, Die Phasenregel, 56.
[2] Meyerhoffer, Zeit. phys. Chem. **5**, 103 (1890).

salt is greater than in the solution in equilibrium with it. The only difference between this and the last subhead is that the solutions containing less of A than the second double salt concentrate to the solution in equilibrium with the first double salt and salt B. No instance of this has yet been found.

d. Both double salts are decomposed by water, an excess of A going into solution in both instances. All solutions concentrate to the second double salt and salt A. This happens at 60° when A = $CuCl_2$ $2H_2O$, B = KCl,[1] and at 80° when A = $MgSO_4 6H_2O$ and B = K_2SO_4.[2] With this latter salt pair the lines for the two double salts will appear as one continuous line, since the ratio of the salt components is the same in the two double salts.

e. Both double salts are decomposed by water, the first with precipitation of B and the second with precipitation of A. All solutions concentrate to the solution in equilibrium with the two double salts. No instance of this has yet been studied. It is impossible that two double salts should be decomposed by water, the first with precipitation of A and the second with precipitation of B. This can be seen by trying to construct a diagram for such a state of things.

Case VI. The solid phases are one of the salts, a double salt and a series of solid solutions. There are five possibilities if it is assumed that the line for the solutions in equilibrium with the solid solutions does not cut the dotted diagonal; otherwise the number is much larger.

a. The double salt is not decomposed by water, and the concentration of A in the liquid solutions is greater than in the solid solutions which separate from them. Solutions containing less of A than the double salt concentrate to the solution in equilibrium with the double salt and salt B; the solution corresponding in composition to the double salt evaporates to dryness without change, while solutions containing more of A concentrate to pure A. No instances of this are known.

b. The double salt is not decomposed by water, and the concentration of A in the liquid solutions is less than in the solid solutions in

---

[1] Meyerhoffer, Zeit. phys. Chem. **5**, 103 (1890).
[2] van der Heide, Ibid. **12**, 416 (1893).

equilibrium with them. This differs from the last in that solutions containing relatively more of A than the double salt concentrate to the solution in equilibrium with double salt and solid solution. No instances are known.

  c. The double salt is decomposed by water with precipitation of A. All solutions concentrate to the solution from which double salt and salt B crystallize simultaneously. This happens when A = NH_4Cl, B = $FeCl_3 6H_2O$.[1]

  d. The double salt is decomposed by water with precipitation of B and the concentration of A is greater in the solutions than in the mix crystals which separate from them. All solutions concentrate to pure A. No instances are known.

  e. The double salt is decomposed by water with precipitation of B, while the concentration of A in the solutions is always less than in the mix crystals in equilibrium with them. All solutions concentrate to double salt and solid solution. No instances are known.

  Case VII. Two series of solid solutions separated by a double salt. Since the double salt is not decomposed by water in any of the known cases it will be simpler to exclude the alternative possibility. There are, then, only three subdivisions to consider.

  a. The concentration of A in the solutions is greater than in the solid solutions which separate from them. Solutions containing less of A than the double salt concentrate to double salt and the solid solutions in which B is solvent; all solutions containing more of A than the double salt concentrate to the double salt and the mix crystals with A as solvent; the solution corresponding to the pure double salt remains unchanged. It seems probable that all the salt pairs giving two series of mix crystals and a double salt come under this head.

  b. The concentration of A in the second series of solid solutions is greater than in the corresponding liquid phases while the reverse is true for the first series of mix crystals. This differs from the last in that solutions containing more of A than the double salt concentrate to pure A.

  c. The line for the first series of solid solutions lies below the

[1] Roozeboom, Zeit. phys. Chem. **10**, 145 (1892).

dotted diagonal and that for the second above. The first set of solutions concentrates to pure B, the second to pure A while the solution corresponding to the double salt evaporates to dryness without change of concentration.

Case VIII. Three series of solid solutions. As not a single system coming under this head has yet been studied, it is not worth while to consider in detail the innumerable possibilities which can arise when no limitations are made. If the assumption is made that no line can cut the dotted diagonal there are but two subdivisions.

a. All the lines lie above the dotted diagonal. All solutions concentrate to pure A.

b. The first line counting from the left lies above the dotted diagonal, the second line lies either above or below it and the third line lies below it. All solutions concentrate to the solution the two points for which lie on opposite sides of the dotted diagonal.

It may be well to state explicitly at the end of this discussion the general rule enabling one to predict from the diagrams the behavior of the different solutions. All solutions on the diagonal or with two points on either side of it evaporate to dryness without change of concentration. All other solutions move to the right with fractional evaporation if the points representing them lie above the diagonal, to the left if below.

If, instead of studying the successive crops of crystals from a given solution, one should recrystallize the solid phases discarding the mother liquor, the order of change would be reversed. The separation of the rare earths by fractional crystallization rests upon the principles discussed in the last two chapters though the problem is usually complicated by there being more than three components.

CHAPTER XVII.

TWO VOLATILE COMPONENTS

Most of the work on systems with three components forming only one liquid phase has been confined to two salts and water owing to the fewer experimental difficulties attending such researches. To Bakhuis Roozeboom we are indebted for a careful study of the system made up of ferric chloride, hydrochloric acid and water.[1] One of the components is a gas at ordinary temperatures and pressures and the system is also interesting because there is an unusually large number of binary compounds possible and all have true melting points. The results of the investigation are expressed graphically in Fig. 53 where the reacting weight of ferric chloride is assumed to correspond to the formula $FeCl_3$. This was not done in considering the equilibrium between ferric chloride and water; but it is thought advisable here in order to make the scale of the diagram more satisfactory. The field for ice as solid phase is bounded by AG'A', for ferric chloride with twelve of water by AG'YVLC; for ferric chloride with seven of water by CLME; for ferric chloride with five of water by EMNG; for ferric chloride with four of water by GNSOJ while anhydrous ferric chloride exists to the right of JOZ. At the points B, D, F and H the solutions have the same concentrations as the hydrates with twelve, seven, five and four of water respectively. The temperatures of the points along this axis are $A - 55°$, $B + 37°$, $C\ 27.4°$, $D\ 32.5°$, $E\ 30°$, $F\ 56°$, $G\ 55°$, $H\ 73.5°$ and $J\ 66°$. Along the hydrochloric acid side of the triangle we have as solid phases the trihydrate bounded by A'G'YWH'C'; the dihydrate bounded by C'H'K'E' and the monohydrate existing in the field above E'K'T'L'. At B' and at D' the solution has the same composition as the hydrate which exists as solid phase[2]. The temperatures of the points on

---

[1] Roozeboom and Schreinemakers, Arch. néerl. **29**, 95 (1894); Zeit. phys. Chem. **15**, 588 (1894).

[2] There must be a misprint in Roozeboom's figures for the concentration at C', for the trihydrate of hydrochloric acid has a true fusion point. The triangular diagram of Roozeboom, Zeit. phys. Chem. **15**, 626 (1894), is in accordance with this fact.

FIG. 53.

this axis are $A' - 90°$, $B' - 25°$, $C' - 35°$, $D' - 17°.5$ and $E' - 20°$. The data for the temperatures are taken from Pickering's measurements[1] and the curves AG', A'G', G'Y, YW, WH', C'H', H'K', E'K', K'T' and TL' are hypothetical ones put in approximately to show the general relations. There have been three ternary compounds studied, the compositions of which may be expressed by the formulas $Fe_2Cl_6 2HCl 12H_2O$, $Fe_2Cl_6 2HCl 8H_2O$, and $Fe_2Cl_6 4H_2O$. The first exists as solid phase in the field VYWV; the second in the field VWH'K'TSNMLV while the field for the third is bounded on one side by L'TSOZ and the other limits have not been determined owing to the difficulties involved in experimenting in sealed tubes. The points W, Y, V, L, M, N, S and O represent the concentrations of nonvariant systems which have been studied and G', Y, W, H', K' and T the concentrations of nonvariant systems which were only guessed

---

[1] Ber. chem. Ges. Berlin, **26**, 280 (1893).

at. The temperatures at which these systems exist are $V - 13°$, $L - 7.5°$, $M - 7.3°$, $N - 16°$, $S - 27.5°$, and $O + 29°$ while the other values are probably $G' - 100°$, $Y - 60°$, $W - 40°$, $H' - 45°$, $K' - 55°$ and $T - 65°$. The line starting from the lower left hand corner of the triangle and running to the middle of the opposite side is the locus of all solutions in which $FeCl_3$ and HCl are present in equivalent quantities. The point R represents a solution having the same composition as the first ternary compound ; Q is the corresponding point for the second and P for the third. The point R does not lie within the field for the compound $Fe_2Cl_6 2HCl 12H_2O$ and this substance has therefore no stable melting point. Owing to the slowness with which transformations take place in systems containing ferric chloride Roozeboom has been able to determine the melting point of this compound which he finds at $-6°$. The solution is instable with respect to ferric chloride with twelve of water. Since the line from the origin through P does not cut the field for $Fe_2Cl_6 2HCl 12H_2O$, this salt will always be decomposed by water with precipitation of ferric chloride with twelve of water. Matters are very different with the second ternary compound $Fe_2Cl_6 2HCl 8H_2O$. The point Q lies within the field for this salt as solid phase and we have the first instance of a ternary compound stable at its melting point. The temperature of Q is $-3°$. The line QRP enters the field for the second ternary compound at $-10°$ and leaves it at $-26.5°$. Between these temperatures the salt will not be decomposed by water and only one solution is stable at each temperature. Between $-3°$ and $-10°$ the salt is not decomposed by water and there are, at each temperature, two solutions with which the compound can be in stable equilibrium. The solutions between $-3°$ and $-10°$ which contain relatively more water than the crystals are stable saturated solutions; the continuous series existing between $-3°$ and $-26.5°$ are stable supersaturated solutions. The line QRP enters the field for the third compound at $-26.5°$ and from that temperature on this compound is not decomposed by water. Since the right-hand boundaries for this field are not known, the conditions under which the salt will again be decomposed by water cannot be stated. At the point P the composition of the solution is the same as that of the crystals. It is not quite certain whether this point lies to the left or the right of the continuation of the line OZ. If to the left, the compound has a stable melting point ; if to the right an instable one. The temperature of the point P is $+ 45.7°$.

In this system we get an excellent confirmation of the theorem of van Rijn van Alkemade[1] that the intersection of the line connecting the melting points of two solid phases with the boundary curve for those phases is a maximum temperature for that monovariant system. The line from Q to L cuts the line LM at a point not marked on the diagram. The temperature of this point is $- 4.5°$ as against $- 7.5°$ and $- 7.3°$ at L and M. The temperature of the point at which the line QP cuts ST is $26.5°$ while S and T represent temperatures of $- 27°.5$ and about $- 65°$. Here the branch OS is very short extending over a range of only one degree. For the curve SO this branch is practically non-existent since the line PH passes through O. In the same way the line QB passes so close to L that it is impossible to distinguish the two points. The line QF connecting the melting points of $Fe_2Cl_6 2HCl 84H_2O$ and $Fe_2Cl_6 5H_2O$ does not cut the stable part of the boundary curve NM at all and the temperature maximum is therefore an instable one. It has been found to occur at $- 5°$, more than two degrees higher than the temperature of the point $M - 7.3°$. The ternary compound $Fe_2Cl_6 2HCl 12H_2O$ is in a state of instable equilibrium at its melting point R and the line QR does not cut the field for this compound. In spite of this, Roozeboom has succeeded in determining the temperature maximum at $- 10.5°$ while the highest temperature at which the two ternary compounds can be in stable equilibrium is $- 13°$, at the point V. At these "fusion points for two solid phases" the isotherms come in contact externally.[2] It is possible to have an internal contact when at least two of the components are common to the two solid phases. This special equilibrium is realized with $Fe_2Cl_6 2HCl 12H_2O$ and $Fe_2Cl_6 12H_2O$. The instable isotherms for the ternary compound start from R as a centre and lie at first entirely inside the isotherms for the binary compound. The former

---

[1] Zeit. phys. Chem. **II**, 311 (1893).

[2] From the diagram it is clear that WH', H'K' and K'T will each show a maximum temperature somewhere on the curve. These maxima have not been determined.

set spread out faster than the latter and become tangent at the point where the prolongation of BR cuts YV. The temperature at this point is $-12.5°$ while that of the point V is $-13°$. The change at this point is from the ternary compound into the binary compound and solution so that the resulting divariant system is always binary compound, solution and vapor irrespective of the relative amounts of the solid phases originally present. Such points are, therefore, called "points of transformation for two solid phases."[1] The two sets of temperature maxima can be distinguished without a knowledge of the isotherms. In the first case, the maximum lies between the two melting points and in the second case beyond them. It is clear that points of transformation for two solid phases cannot occur when both solid phases have stable melting points.

The forms of the isotherms for $+10°$ and $-10°$ have been marked in dotted lines. The first does not cut the field for either of the first two ternary compounds. It passes through the point R, the temperature at which a solution having this composition can be in equilibrium with $Fe_2Cl_4 12H_2O$ being $+10°$. This same solution can be in instable equilibrium with $Fe_2Cl_2 2HCl 12H_2O$ at $3°$, as we have already seen. The isotherm for $-10°$ cuts the field for $Fe_2Cl_4 2HCl 8H_2O$ passing round the point Q. The isotherm for the intermediate temperature of $-4°$ would be very like that for $+10°$ plus a closed curve round Q. The isotherm for $-6°$ would have the same general form as that for $-10°$ with the addition of an instable closed curve round R. The isotherm for $30°$ makes a disconnected semi-circular figure round B and a curve round D, meeting the boundary curve at E. The isotherm then cuts FX and JO, takes a turn round O, intersecting OZ, and then passes up into the field for $Fe_2Cl_2 2HCl 4H_2O$ and out of the diagram. In all the fields for the binary compounds the ratio of hydrochloric acid to ferric chloride passes through a maximum. A host of other isotherms have been determined by Roozeboom and Schreinemakers; but they present no new features which require discussion here. Since all the solid phases, save one, are stable at the melting points, it is impossible to predict whether a given quintuple point will be a maximum or a minimum temperature for any phase or not.

[1] Roozeboom. Zeit. phys. Chem. **15**, 619 (1893).

In Table XXIX are some of the data for the boundary curves, the concentrations being expressed as reacting weights of hydrochloric acid, and of ferric chloride, $FeCl_3$, in one hundred reacting weights of the solution. Values marked with an asterisk are estimated and not determined directly. In Table XXX are the temperatures and concentrations for the different nonvariant systems. The values for the points G', Y, W, H', K' and T are only approximative.

Table XXIX

Temp.	HCl	$FeCl_3$	Temp.	HCl	$FeCl_3$	Temp.	HCl	$FeCl_3$
Curve CL			Curve JO			Curve NM		
+27.4°	0.0	19.6	+66.°	0.0	36.8	− 7.3°	15.2	18.8
26.5	0.2	19.5	60.	6.4	33.3	− 5.	14.4	18.2
25.	1.9	18.8	55.	9.8	31.5	Curve ML		
20.	4.3	18.3	50.	11.6	30.4	− 7.3	15.2	18.8
15.	8.0	17.5	40.	15.2	28.6	− 4.5	14.1	17.6
10.	10.1	17.0	30.	17.9	27.3	− 7.5	13.5	16.6
0.	12.8	16.6	29.	18.4	27.1	Curve LV		
− 7.5	13.5	16.6	Curve OZ			− 7.5	15.5	16.6
Curve EM			29.	18.4	27.1	−10.	14.5	14.6
+30.	0.0	23.2	30.	18.7	27.0	−13.	16.0	12.8
25.	5.5	21.7	35.	19.9	26.4	Curve VY		
20.	7.9	20.8	40.	21.5	25.5	−13.	16.0	12.8
15.	10.4	19.8	Curve OS			−12.5	16.0	12.0
10.	12.2	19.1	29.	18.4	27.1	−15.	16.3	7.4
− 7.3	15.2	18.8	25.	19.5*	25.4*	−20.	16.1	5.4
Curve GN			20.	19.8	24.2	−60.*	15.5*	3.0*
+55.	0.0	28.9	10.	20.0	22.8	Curve VW		
50.	2.3	28.6	0.	20.1	22.1	−13.	16.0	12.8
44.	7.2	25.6	−10.	19.8	21.3	−15.	17.5	11.3
40.	8.9	24.8	−20.	19.5	21.1	−20.	18.4	6.7
33.	10.3	24.3	−27.5	19.4	20.6	−40.	19.3*	4.6*
30.	11.1	23.9	Curve SN			Curve ST		
25.	12.6	22.8	−27.5	19.4	20.6	−27.5	19.4	20.6
20.	13.7	22.0	−20.	18.5	20.1	−26.5	20.1	20.1
10.	15.6	20.8	−16.	17.7	19.9	−29.5	21.7	21.9
0.	16.4	20.3	Curve NM			−65.5*	29.9*	15.5*
−10.	17.1	20.1	−16.	17.7	19.9			
−16.	17.7	19.9	−10.	16.6	19.5			

Table XXX

	Temp.	HCl	FeCl$_2$		Temp.	HCl	FeCl$_2$
V	−13.°	16.0	12.8	G'	−100°	12.9	12.4
L	− 7.5	13.5	16.6	Y	− 60	15.5	3.0
M	− 7.3	15.2	18.8	W	− 40	19.3	4.6
N	−16.	17.7	19.9	H'	− 45	24.5	4.9
S	−27.5	19.4	20.6	K'	− 55	29.6	8.6
O	+29.	18.4	27.1	T	− 65	29.9	15.5

CHAPTER XVIIII

COMPONENTS AND CONSTITUENTS

In none of the systems considered thus far has there been any difficulty in determining the number and nature of the components; but this is not always the case. When two compounds can react to form two others by metathesis, it is not obvious how many components there are, while matters are complicated still more by the possibility of several sets of components. This makes it necessary to discuss the question of the number and choice of components.[1] It must be kept in mind that there are $n + 2$ phases in a nonvariant system only when the n components are independently variable. Gibbs[2] starts from the fact that a system containing n independently variable components and r phases is capable of $n + 2 - r$ independent variations when we exclude "passive resistances to change, effects due to gravity, electricity, distortion of the solid masses and capillary tensions." From this it follows that a system with n independently variable components is completely defined when there are $n + 2$ coexisting phases. By reversing the argument one can deduce from the number of possible coexisting phases the numbers of independent variables; but nothing about the total number of components, for the very simple reason that there is no limit to the number which may be selected. Gibbs[3] states that there may be any number of components, of which certain ones are dependently variable and have no influence on the number of possible coexisting phases. It will simplify matters to introduce a new term and to call the arbitrarily chosen substances, from which a given system can be formed, the "constituents"[4] of that system. We can then use the

---

[1] Cf. also Roozeboom, Zeit. phys. Chem. **15**, 150 (1894); Wald, Ibid. **18**, 337 (1895).
[2] Trans. Conn. Acad. **3**, 152 (1876).
[3] Ibid. **3**, 124 (1876).
[4] Cf. Trevor, Jour. Phys. Chem. **1**, 22 (1896).

word "**component**" to denote the substances in the system which are capable of independent variation. For a given system there can be different sets of constituents. To take an extremely simple case, the constituents of **ammonium** chloride might be ammonium chloride; ammonia and hydrochloric acid; the two radicles, ammonium and chlorine; or **nitrogen, hydrogen and chlorine**, just as seemed best under the circumstances. The number of constituents may equal, exceed, or be less than the number of components. If there are N constituents and h relations among them, the number of components n will be $N - h$. Stated in words, each relation among the constituents reduces the number of components by one. This way of looking at the problem simplifies matters because we can take as constituents the chemical **elements, or any groups of elements, and the** difference between the sum of these **artificially chosen constituents** and the number of limiting **conditions gives the number of components**.

A few illustrations will make this clear. In a system containing hydrogen and oxygen, we may say that there are two constituents. If there are no **limitations, both of these are independently variable and there are two components.** If the ratio by weight of hydrogen to oxygen is one to eight in every phase, we have introduced the limitation :

$$2H = O,$$

and there is but one component, commonly **called water.** A system containing potassium, nitrogen and oxygen, will have only one component, potassium nitrate, if the **two limitations hold for all the** phases :

$$K = N = 3O.$$

If we take K, O and H as the three constituents **of a certain system, there will be one component if there are the two limitations :**

$$K = O = H.$$

There will be two components **if there is the one limitation :**

$$K + H = OH.$$

In this latter case the only solid phases possible will be ice, potassium hydroxide and potassium hydroxide with varying amounts of water of crystallization. If metallic **potassium,** potassium oxide, hydrogen or oxygen is a possible phase under the conditions of the experiment, the limiting condition does not hold and there are three components. If K, NO_3 and Cl are taken as constituents in a system composed of potassium nitrate and potassium chloride, there is one limitation, namely, that for every phase :

$$K = NO_3 + Cl.$$

If the system had been supposed to be made up of the chemical elements, there would have been four constituents instead of three ; but there would have been two limiting conditions instead of one, so that, from either point of view, the number of components is two. If potassium nitrate and potassium chloride had been taken as the constituents there would have been no limiting conditions.

If we say that the system, potassium nitrate, potassium chloride and potassium bromide, contains four constituents, K, NO_3, Cl and Br, there is one limiting condition :

$$K = NO_3 + Cl + Br.$$

In the same way systems made from potassium nitrate and sodium chloride may be said to have four constituents, K, NO_3, Na and Cl ; but there are only three components because of the limitation :

$$K + Na = NO_3 + Cl.$$

If the two salts are taken in equivalent quantities, there are only two components, the extra limiting condition being :

$$K = NO_3 \text{ or } Na = Cl.$$

Potassium chloride and water contain four elements but only two components, if we insist that in all the phases there are the relations :

$$K = Cl \text{ and } 2H = O.$$

If the salt is decomposed by water and the water acts as a monobasic acid there will be only the one limitation :

$$K + H = Cl + OH.$$

If the water can act also as a dibasic acid there are no limiting conditions and the number of components is four. This state of things is difficult to realize at any convenient temperature with potassium chloride and water, and it will be more satisfactory to consider the action of water upon lead chloride. At temperatures at which there is no decomposition, the number of components is two, for there are two limiting conditions which may be expressed in two ways:

$$Pb = 2Cl \text{ and } 2H = O,$$
$$\text{or, } PbCl = Cl \text{ and } H = OH.$$

If lead oxide hydrochloride, PbOHCl, can be formed under the conditions of the experiment, there will be three components since there is only the one relation:

$$PbCl + H = Cl + OH.$$

If lead oxide can also be formed there will be four components since there are then no limiting conditions. If lead oxide is a possible phase and lead oxide hydrochloride is not, there will be three components, the single relation being:

$$Pb + 2H = O + 2Cl.$$

This same limiting condition would seem to hold throughout if we we were to double the formula for lead oxide hydrochloride, and we are thus confronted with the difficulty that a system seems to contain four components if the simplest formula for one of the compounds be taken and three components if the formula be doubled. This would imply that the number of components could be determined only after we knew the reacting weights of all the constituents or, taking it the other way round, that the reacting weights could be determined from a knowledge of the variance of the system. The fallacy here is due to our forgetting that a formula is merely a concise way of stating known facts. If we write the doubled formula of lead oxide hydrochloride in the form (PbOHCl)$_2$, we imply that the compound is an addition product of the—possibly hypothetical—substance PbOHCl. If we write the formula PbO$_2$H$_2$PbCl$_2$ or PbOPbCl$_2$H$_2$O we imply that the compound is an addition product of lead hydrate and lead chloride, or a hydrate of the oxide and chloride of lead. If the first view is correct the system will contain four components, because there is no relation among the constituents; if either of the other views is correct, there are only three components because we have the relation:

$$Pb + 2H = O + 2Cl.$$

In the first case the substance will dissociate with formation of hydrochloric acid; in the others with formation of water. Experimentally, the first reaction is the one that occurs. This same point was involved in the discussion of the equilibrium for K, O and H on page 227, though it was not gone into there. It was shown that there were three components if potassium oxide was a possible phase. A case where water is only a dibasic acid is to be found with mercuric sulfate and water. The number of components is therefore three, and if we take Hg, SO_4, H_2 and O as constituents, we have one limiting condition :

$$Hg + 2H = SO_4 + O.$$

If we take HgO, SO_3 and H_2O as constituents, there are no limiting conditions and the number of constituents equals the number of components. A system made up of copper sulfate, potassium sulfate and water contains three components, under ordinary circumstances. The number of components will be reduced to two if in each phase there exists the relation :

$$CuSO_4 = K_2SO_4.$$

If the temperature is raised until copper sulfate is attacked by water there will be four components. Under these circumstances the number of components becomes greater than the number of constituents arbitrarily selected and therefore h must be negative. Since the positive value of h denotes a limiting condition, a negative value must denote a condition of freedom. Adding the assumption that copper sulfate reacts with water is subtracting a limiting condition or adding a degree of freedom. This can be seen more clearly if we take potassium, copper, hydrogen, sulfur and oxygen as constituents, five in all. If neither of the salts is decomposed by water there are two limitations :

$$2K + Cu = SO_4,$$
$$2H + 2K + Cu = SO_4 + O,$$

and therefore three components. If the water can decompose the salts the first limitation drops out and there are four components. The system, made from potassium nitrate and sodium chloride, has already been considered on the assumption that the radicles are the constituents. It is equally justifiable to take the two salts as the constituents ; and then

$$h = -1,$$

because the two salts may react with each other giving an extra degree of freedom. The number of components is therefore three, which is the value found before, as of course it must be. While it is convenient to take the known chemical elements or groups of these elements as the constituents, this is not necessary, and the Phase Rule does not rest in any way upon the permanency of the present chemical elements. If copper should be separated into two new elements, for instance, these could be taken as constituents; but, in the systems in which there occurs the substance which we now call copper, there would be the limiting condition that the two new elements must be present in the proportions in which they formed copper. In the same way, while there may be an increase in the number of constituents in a given phase, owing to what we call electrolytic dissociation, there is no change in the number of components.

The system composed of potassium nitrate and chloride presents several characteristics which have not been met before. It was first pointed out by Meyerhoffer[1] that it contained three components, and not two or four as might readily have been supposed. He also called attention to the fact that it is impossible to select three substances and to say that these and these only are the three components. It is absolutely immaterial whether one takes potassium chloride, potassium nitrate and sodium chloride; potassium chloride, potassium nitrate and sodium nitrate; potassium chloride, sodium chloride and sodium nitrate; or potassium nitrate, sodium chloride and sodium nitrate as the components. It is impossible to use the ordinary triangular diagram to represent the states of equilibrium. If three of the salts occupy the three corners of the triangle, there is no place in the diagram for the fourth salt, although the concentration of all possible solutions can of course be represented. Roozeboom[2] has suggested the use of a double diagram consisting of two equilateral triangles placed with bases together so as to form a lozenge, the four salts occupying the four corners. The concentrations of the solutions may then be expressed in either triangle, those of the solid phases in one or the other but not both of the triangles. This leaves the vertical axis for the temperature as before. If one does not care about

---

[1] Die Phasenregel, 80.
[2] Zeit. phys. Chem. **15**, 155 (1894).

expressing the temperature but is willing to use a solid figure, the natural form of the diagram would be a tetrahedron.

No system of this general type has yet been studied; but since three solid phases, solution and vapor constitute a nonvariant system, the monovariant system which is formed by withdrawal of heat will be three solid phases and vapor. From this it follows that there is one pair of salts which can not exist in stable equilibrium with solution and vapor. For this reason, in using Roozeboom's diagram, it is necessary to place the two salts which can not coexist with solution and vapor at the opposite ends of the long diagonal. There may be another inversion point at some lower temperature at which the four salts are in equilibrium with each other and with vapor. No such case has yet been observed, though Spring[1] claims to have shown the existence of the monovariant system, sodium sulfate, barium carbonate, barium sulfate and sodium carbonate, at pressures up to six thousand atmospheres—a statement which must be erroneous.

Mercuric sulfate and water react with formation of mercuric oxide and sulfuric acid. If we consider these four substances as the constituents of a system there will be one limiting condition and three components. With most cases of metathesis, that is the only convenient way of treating the system; but, in this particular instance, it is possible to take mercuric oxide, sulfur trioxide and water as constituents, and there are then no limiting conditions. Hoitsema[2] has considered the system from this point of view, and it is a most excellent one for this particular case; but he makes the mistake of suggesting that all systems containing basic salts should be treated in the same way, which would neither be general nor convenient. When sulfuric acid is added continuously to mixtures of mercuric oxide and water, the solid phases which appear successively at $25°$ are $HgSO_4 2HgO$, $HgO_2 HgSO_4 2H_2O$, $HgSO_4 H_2O$ and $HgSO_4$, while at $50°$ the phase $HgSO_4 H_2O$ no longer appears.

In a series of studies upon the alkaline tartrates van 't Hoff[3] has introduced certain limitations which convert a system really composed

---

[1] Bull. soc. chim. **46**, 299 (1886).
[2] Zeit. phys. Chem. **17**, 651 (1895).
[3] Ibid. **1**, 173 (1887); **17**, 49, 505 (1895).

of four components into one made up of three. If we take the sodium salt of dextrorotary tartaric acid and the ammonium salt of laevorotary tartaric acid, we shall have three components, for the same reason that potassium nitrate and sodium chloride form a three-component system. Adding water we shall have four components, and it will require six phases to make a nonvariant system. It is found experimentally that, under the conditions of the experiment, there is an inversion point at 27° where the solid phases in equilibrium with solution and vapor are the dextrorotary sodium ammonium tartrate, the laevorotary sodium ammonium tartrate, and the sodium ammonium racemate. There are only five phases and therefore a limiting condition must have been introduced. This is the case, for it is an essential part of the experiment that one starts with sodium ammonium racemate, and the limiting condition is therefore:

$$Na = NH_4.$$

Another inversion point occurs at 35°, and here the solid phases are the sodium ammonium racemate, sodium racemate and ammonium racemate, the limiting condition being:

$$l\text{-}C_4H_4O_6 = d\text{-}C_4H_4O_6.$$

The two limiting conditions are different in the two cases and it might be thought that it would not be possible to observe the two inversion points with the same original mixture. It is to be noticed that both these conditions are satisfied in the sodium ammonium racemate, and therefore if we add water to this salt and follow the changes with the temperature we shall find two inversion points, one at 27° and the other at 35°. Starting with any other mixture of the sodium ammonium tartrates and water, we should find an inversion point at 27° but none at 35°. If we should start with any mixture of sodium racemate, ammonium racemate and water, not containing the salts in equivalent quantities, there would be an inversion point at 35° but none at 27°. By varying the conditions it is thus possible to make the system behave as if it contained three components at all temperatures, three components below a certain temperature and four above it, or three components above and four below another definite temperature. No experiments have yet been made with the system having four components at all temperatures. This would be an interesting matter to work upon because it is very probable that each inversion temperature would correspond to two nonvariant systems, owing to the dextrorotary and the laevorotary salts having the same solubility. There would, of course, be four solid phases in equilibrium with solution and vapor at these inversion points.

Since each limiting condition reduces the number of components by one, it is clear that if for any reason a limiting condition becomes inoperative the number of components will be increased by one. Such a state of things occurs whenever there is a passive resistance to change. Whenever an expected reaction does not take place, it is said that there is a passive resistance to change. This is a definition and not an explanation. In some cases the system is in a state of labile equilibrium. In others it appears to be in a state of stable equilibrium, but the accepted doctrine is that this is an instance of an immeasurably low reaction velocity, and that the system is not in equilibrium at all.[1] If we bring in the element of time, it is possible to apply the Phase Rule to these cases. If a reaction does not take place to a measurable extent within the time under consideration, it has no effect as a limiting condition for that time. For instance, a mixture of hydrogen, oxygen and water at ordinary temperature is to be considered as having three components if the experiment does not last over a month because, for that time, there is no such reaction as:

$$2H_2 + O_2 = 2H_2O.$$

It is quite possible, if the experiment lasted a thousand years, that the system would have to be treated as containing two components, though this is not proved.

When the limiting condition takes the form of a chemical reaction, its effect is to change the number of components; but it is possible to introduce conditions which will affect the number of pressures. This can be done by dividing the system into two parts by a diaphragm permeable to one of the components and impermeable to the others.[2] For the sake of simplicity we will consider a system of two components, and if we have in one compartment a solution and in the

---

[1] Nernst, Theor. Chem. 340.
[2] Gibbs, Trans. Conn. Acad. **3**, 138 (1876).

other the component which can pass through the diaphragm, a moment's consideration will show that there can not be equilibrium when the pressures on the two sides of the diaphragm are equal. The concentrations in the two liquid phases are not equal, nor in the two vapor phases which could be in equilibrium with them. There will therefore be a tendency for one component to pass through the diaphragm diffusing from the pure liquid into the solution, and this diffusion can be checked only by increased pressure upon the solution phase. Our conclusion that $n + 2$ phases constitute a nonvariant system was based on the assumption that the pressure and temperature were uniform throughout the system, or that the number of independent variables equalled the n components plus the temperature and pressure. If there are two independently variable pressures, the total number of independent variables is $n + 3$, and it will take this number of phases to constitute a nonvariant system while $n + 2$ and $n + 1$ phases will be necessary for monovariant and divariant systems respectively. Before considering such systems it will be well to ask whether a diaphragm is possible which shall be permeable to one component and impermeable to the other. Experiment shows that such a thing can exist, and diaphragms of this sort are usually called semipermeable,—a barbarous term which has been universally adopted. Traube[1] showed that when solutions of two substances were brought together carefully there was often formed a coherent membrane at the surface separating the two solutions, provided the two substances reacted to give a colloidal precipitate. These membranes were always permeable to water; but often impermeable to dissolved substances. One of the best of these membranes is made of copper ferrocyanide, and this seems to be impermeable to more substances than any other known. An extended study of the permeability of different membranes is to be found in a paper by Walden.[2] Pfeffer[3] succeeded in giving these membranes the strength to withstand pressure by precipitating them in the walls of a porous cell. This is very difficult to do successfully, and Pfeffer alone has succeeded in making satisfactory measurements. Semipermeable

---

[1] Cf. Ostwald, Lehrbuch I, 653.
[2] Zeit. phys. Chem. 10, 699 (1892).
[3] Osmotische Untersuchungen, Leipzig (1877).

membranes occur very largely in nature, having been found in plant cells, blood corpuscles, bacteria, and elsewhere;[1] but these have not yet been used to measure directly the pressure caused by the tendency of the pure liquid to flow into the solution. This pressure is commonly called the osmotic pressure. Pfeffer measured the pressure necessary to keep the solution in equilibrium with water. The system which he studied consisted of three phases, solution, solvent and vapor. For a case of this sort we have seen that $n + 1$ phases constitute a divariant system and that it is necessary to fix two variables before the system is completely defined. Pfeffer found that for each temperature there could be a series of pressures and for each pressure a series of temperatures, depending on the concentration of the solution phase. If the temperature and concentration be fixed there is but one osmotic pressure at which the system can be in equilibrium. The experiment of changing the pressure upon the water outside was not tried, but there is no doubt that as soon as the vapor phase had disappeared, the osmotic pressure would have varied with varying external pressure, the temperature and the concentration being kept constant. Very recently Raoult[2] published a note to the effect that vulcanized rubber is permeable to ether and impermeable to methyl alcohol, while pig's bladder is permeable to methyl alcohol and impermeable to ether. If we imagine methyl alcohol and its vapor separated from a solution of methyl alcohol and ether by a diaphragm of pig's bladder, and this same solution separated from ether and ether vapor by a diaphragm of vulcanized rubber, there will be a tendency for methyl alcohol to flow into the solution from one side which can be neutralized by a pressure P_1 upon the solution. There will be a tendency for ether to flow in from the other side which can be neutralized by a pressure P_2 upon the solution. Since the solution can not be under two different pressures, one or the other of the liquids will flow out of the cell until $P_1 = P_2$; *i. e.*, until the osmotic concentrations of methyl alcohol and ether in the solution are the same. In other words, the final equilibrium is independent of the

---

[1] Cf. Zeit. phys. Chem. **2**, 415 (1888); **3**, 103 (1889); **6**, 319 (1890); **7**, 529; **8**, 685 (1891); **16**, 261 (1895).
[2] Ibid. **17**, 737 (1895).

initial concentration in the cell.[1] This conclusion can be reached more satisfactorily by an application of the Phase Rule. There are five phases, liquid methyl alcohol, vapor of methyl alcohol, solution, liquid ether and vapor of ether. There are two diaphragms and consequently two extra pressures, so that $n + 4$ phases constitute a nonvariant system and $n + 3$ a monovariant one. The number of components being two, we are discussing a monovariant system and it follows that for a given temperature there can be only one concentration in the solution phase for which the system is in equilibrium.

Returning to the simpler case with only one diaphragm and assuming that the external pressure upon the solution is not great enough to prevent the formation of a vapor space, it is clear that all the liquid will diffuse in from outside and, at equilibrium, there will be solution, vapor above the solution and the vapor outside the cell, which of course will not be saturated vapor. At temperatures below the freezing point it would be possible to have ice and vapor in equilibrium with the solution.

There is no theoretical reason why there should not be systems with two temperatures instead of two pressures, if we could find diaphragms permeable to the components and impermeable to heat. Since no diaphragm is known which is impermeable to heat, it is not worth while to discuss what would happen in case this hypothetical membrane were permeable to the components.

---

[1] My attention was called to this case by Mr. D. McIntosh.

CHAPTER XIX

TWO LIQUID PHASES

No systems have yet been studied in detail in which there are three components and two liquid phases. If one starts at the temperature T with two components, A and B, forming the system, solid A, two solutions and vapor, and adds a third component, C, there will be formed a monovariant system which can exist over a range of temperature limited only by the appearance of a new phase or the disappearance of an old one. When the third component is a solid at all the temperatures under consideration the effect of adding it in excess will usually be to form the nonvariant system, solid A, solid C, two solutions and vapor. If the temperature T is very near the fusion point of pure B, it will be possible by changing the nature of C to have nonvariant systems of this type formed in which the solid phases are A and C, A and B, or B and C. This could be realized with phenol, water and a third component or benzene, water and a third component. Let benzene be represented by A, water by B, and the third component by C. If C is a substance soluble in water and very sparingly soluble in benzene, the solid phases will be A and C. If C is soluble in benzene and very sparingly soluble in water, the solid phases will be A and B, when C is not present in excess or B and C when it is. It is to be noticed that the temperatures at which these nonvariant systems exist are not necessarily lower than those of the binary nonvariant systems from which they are derived. If the component C is soluble in A and not soluble in B the equilibrium temperature will be lowered by addition of C. If C is soluble in B and practically insoluble in A the temperature will rise, the amount varying with the nature of C and the mutual solubility of A and B. When the third component is a liquid at all temperatures under consideration, these are four important cases. The liquids B and C may be consolute, while the liquids A and C are sparingly miscible. A nonvariant system with two liquid phases is impossible. An example of this would be found when A denotes water, B ether and C chloroform.

The liquids **A** and C may be consolute while the liquids B and C are sparingly **miscible**. The **nonvariant** system will have A and B as solid phases. This can be realized when A is benzene, B is water and **C is** chloroform or ether. The liquid C may be miscible in all proportions with the liquids A and B. While **there** are no experimental data upon which to base an **opinion, it** seems very improbable **that** a nonvariant system can be formed **with the phases A, B,** two solutions and vapor, since the two **solutions** approach **each** other in composition with increasing addition **of C, and it is to** be expected **that the two** will become identical **in composition and cease to exist as separate** phases before B separates as **solid phase. An example of this** furnished **by the** system, naphthalene, **water and alcohol. The** component C may not be miscible in all proportions **with either of the** liquids A and B. Under these circumstances it would **be possible to** have formed a nonvariant system **made up of three liquid phases, solid and vapor.** This can be realized with **sulfur, toluene and water,** though the temperature and pressure **under which this system can** exist have not been determined.

The two solid phases which are in equilibrium with **the two solutions and vapor at the quintuple point may be the pure components** or compounds **or** solid solutions. **By** adding ether to a **solution of** calcium chloride it would be possible **to determine several inversion** points at which the solid phases should be two **compounds. The addition of water to a mixture of benzene and iodine would give a nonvariant** system made up of ice, a solid **solution of iodine in benzene, two** liquid phases **and** vapor, provided **the iodine were present in small** quantities **only. In course of time cases will be found where the two** solid phases are two sets of solid solutions, though no instance of **this** has yet been studied qualitatively.

There are no data for nonvariant **or for monovariant systems** containing two liquid phases. On **the other** hand, divariant **systems** composed of two liquid phases and vapor have received a great deal of attention, though the question has not been taken up with the view of applying the Phase Rule.[1] For this reason most of the experiments have been made with dilute solutions only, and the con-

---

[1] Berthelot and Jungfleisch, Ann. chim. phys. (4) **26**, 396 (1872); Nernst. Zeit. phys. Chem. **8**, 110 (1891); Jakowkin, Ibid. **18**, 585 (1895), *etc., etc.*

centration of but one component has been determined. These systems show the characteristics of divariant systems. If the temperature alone is fixed, the concentrations in the three phases can vary. If, in addition, the composition of one phase is fixed, the concentrations of the other phases are no longer variable. There are a few measurements giving the composition of one of the liquid phases.[1] Although the systems studied contained only two phases, liquid and vapor, yet it was the limiting liquid phase which was investigated, which is, of course, the one which can exist in equilibrium with a second liquid phase. For this reason the experiments referred to are really studies of divariant systems. It was found that the curves for the two solutions met at an angle at the point where the two liquid layers became identical in composition. This result is interesting as a confirmation of the assumption that the same phenomenon often occurs when two liquids become consolute. In the latter case it is very difficult to determine whether the apparent intersection of the lines is real or due to experimental error. Where two liquid phases containing three components become consolute, there can be no question about the facts.

So far it has been assumed explicitly that two of the components could form two liquid layers at some temperature under consideration, and that the third component merely displaced the equilibrium to a certain extent. There are cases known which do not come under this head; where no two of the components form two liquid layers at the temperature of the experiment, and this particular equilibrium can occur only when the three components are taken together in the proper proportions. Alcohol and water are miscible in all proportions at ordinary temperatures; but the addition of certain salts will cause the appearance of two liquid layers, although these salts can not form two liquid phases with water alone or with alcohol alone at the same temperature. Some of the salts which will do this are potassium and sodium hydroxides, sodium phosphate, potassium carbonate, sodium carbonate, ammonium sulfate, sodium sulfate, magnesium sulfate, manganese sulfate, strontium chloride and many others. The same phenomena occur with water and lactones in presence

[1] Pfeiffer, Ibid. **9**, 444 (1892); Bancroft, Phys. Rev. **3**, 21 (1895).

of alkaline carbonates, while most salts have the power of precipitating acetone from aqueous solution at some temperature. In all cases yet known the tendency to form two liquid layers when a third component is added to two consolute liquids increases with rising temperature, solutions which are homogeneous at one temperature separating into two liquid phases when heated. At yet higher temperatures it is probable that the two solution phases would again cease to exist though there are no experiments to show it.

The system, ammonium sulfate, water and alcohol, has been studied by Traube and Neuberg[1] and by Bodländer;[2] but the measurements cover too little ground to permit of making a complete diagram. They are sufficient to enable us to predict the general form of the isotherm at different temperatures. At low temperatures the only divariant system possible is that of salt, solution and vapor, the curve having a break at the point where the solvent changes from water to alcohol. At higher temperatures these two portions of the curve are separated by the curves for the divariant system, two solutions and vapor. If alcohol be added continuously to a saturated aqueous solution of ammonium sulfate, the system does not pass through this last stage. There will first be a precipitation of salt till a definite concentration is reached. There will then appear a second liquid phase forming the monovariant system, salt, two solutions and vapor. With further addition of alcohol there will be increased precipitation of salt and a change in the quantities though not in the concentrations of the two liquid phases. When one of the liquid phases has disappeared completely, the concentration of the remaining solution phase will change with addition of alcohol.

In many instances where the second liquid phase is instable it may yet be formed temporarily. When alcohol is added to a strong solution of sodium carbonate, at ordinary temperature, the liquid usually separates into two layers, one of which is instable and disappears in the course of time. This phenomenon seems to be entirely general and to occur in all cases when a solid is precipitated from a solution by addition of a liquid in which it is not soluble. This was first shown by Link and by Schmidt, and the subject has since been

---

[1] Zeit. phys. Chem. **1**, 509 (1887). [2] Ibid. **7**, 308 (1891).

studied by many.[1] Frankenheim found that such salts as ammonium chloride, magnesium chromate, calcium carbonate and the sulfates of sodium, magnesium, manganese and aluminum separate in liquid drops when precipitated by alcohol. From this it is fair to conclude that all salts are precipitated as an instable solution from which the solid then crystallizes. This raises the question whether a solution of two components may not usually pass through the instable state of two solutions before crystallizing, and this receives a certain confirmation from the fact that with salicylic acid and water the curve for two solutions and vapor is instable along its whole length.[2] It is impossible to answer this question directly, and so far no direct method of doing this has been discovered.

[1] Ostwald, Lehrbuch I, 1040; Lehmann, Molekularphysik I, 730.
[2] Cf. also Lehmann, Molekularphysik I, 726.

FOUR COMPONENTS

CHAPTER XX

GENERAL THEORY

With four components, six phases constitute a nonvariant system, five a monovariant system, and four a divariant system. The discussion will be limited to cases in which water is one of the components, because no other system has yet received any attention. When the other three components are three salts, such as the chloride, bromide and iodide of potassium, which do not react with each other and which form no hydrates, the only nonvariant system possible has the three salts and ice as solid phases. In the monovariant systems in which solution and vapor occur, the other phases can be the three salts or any two of the salts and ice. When double salts or hydrates are possible, the number of nonvariant systems will be correspondingly increased. When the constituents are water and two salts, A and B, which can form two others, C and D, by metathesis, there are four components. If there is a sextuple point in which the four salts are in equilibrium with solution and vapor there will be two boundary curves for three salts, solution and vapor existing at temperatures above the sextuple point and two others existing at temperatures below it. If the solid phase, along the latter pair of curves are A, B, C and A, B, D, respectively, the solid phases at higher temperatures will be A, C, D and B, C, D, the sextuple point being an inversion point for the two pairs of salt, A and B, C and D. When the solid phases are the anhydrous salts such a sextuple point can occur when the salts by themselves can form the nonvariant system, four salts and vapor. Since no instance of this has yet been found, no sextuple point of this type has been discovered. If any mixture of four salts, which form no compounds with water, be dissolved in water and the solution evaporated to dryness, one pair of salts will be found to be stable and the other pair instable. Instances of stable pairs of salts are potassium nitrate and sodium chloride, sodium nitrate and ammonium chloride. The reciprocal pairs, potassium chloride and sodium nitrate, sodium chloride and ammonium nitrate, are not stable and will not crystallize from solution simultaneously. While it is possible to have potassium nitrate, sodium chloride and sodium nitrate or potassium nitrate, sodium chloride and potassium chloride in equilibrium with solution and vapor, it does not follow that this equilibrium will be reached unless three out of the four salts be added to the solution. If one starts with potassium nitrate, sodium chloride and water, this is really a system containing three components, because the two salts do not react and the system differs experimentally in no way from one made up of potassium nitrate, potassium chloride and water. If we take K, Na, Cl, NO_3 and H_2O as the constituents there are two limiting conditions:

$$K = NO_3 \text{ and } Na = Cl.$$

This is true only so long as the experiments are carried on at temperatures at which potassium nitrate and sodium chloride are the stable pair. Beyond the inversion temperature these two limiting conditions would be replaced by the single one:

$$K + Na = NO_3 + Cl.$$

This can be realized experimentally by starting with the instable pair of salts. If we dissolve potassium chloride and sodium nitrate in water and evaporate to dryness, the solid phases will be potassium nitrate, sodium chloride and either potassium chloride or sodium nitrate, depending on which salt is present in excess. If the two salts are taken in equivalent quantities there will be no excess and the solid phases are potassium nitrate and sodium chloride. Under these circumstances the system behaves as if it contained only three components; but this case is not to be confused with the one where the salt constituents are the stable pair. If one starts with the stable pair of salts and water there are three components until the inversion temperature is passed when the number increases to four. If one starts with the instable pair and water, the two salts being in equivalent quantities, or with the stable pair and water, the two salts being in equivalent quantities, the system contains three components whether the experiments be carried on above or below the inversion temperature. At one side of the sextuple point the limiting conditions are:

$$K = NO_3 \text{ and } Na = Cl,$$

while on the other side of the point we have:

$$K = Cl \text{ and } Na = NO_3.$$

It is clear, of course, that one could write the limiting conditions somewhat differently if desired as, for instance,

$$K + Na = NO_3 + Cl \text{ and } K = NO_3$$
$$K + Na = NO_3 + Cl \text{ and } K = Cl.$$

The one set applies when potassium nitrate and sodium chloride are the stable pair and the other in the reverse case. This is merely an algebraical transformation; but it has the advantage of simplicity in that only one condition changes as the system passes through the inversion point.

When at least one of the salts contains water of crystallization, it is not difficult to find instances of sextuple points in which ice is not one of the solid phases. The study of these points has been carried on entirely in van 't Hoff's laboratory. At $3.7°$ there can be in equilibrium hydrated sodium sulfate, potassium chloride, sodium chloride, a double sulfate corresponding to the formula $K_2Na_4SO_4 \cdot_4$, solution and vapor. The first pair of salts is stable below this temperature, the second above it.[1] At $10.8°$ ammonium chloride and hydrated sodium sulfate react to form sodium chloride and sodium ammonium sulfate with two of water, the latter pair being stable above this temperature.[1] At $31°$ sodium chloride and hydrated magnesium sulfate change into sodium magnesium sulfate with four of water and hydrated magnesium chloride, the latter pair being stable above this temperature.[2] By adding sodium chloride in excess the point at which the double sulfate of sodium and magnesium changes into the single sulfates is lowered to $5°$.[3]

It seems probable that in the near future we shall have complete data for at least one system containing four components;[3] but for the

---

[1] van 't Hoff and Reicher, Zeit. phys. Chem. **3**, 482 (1889).
[2] van 't Hoff and van Deventer, Ibid. **1**, 165 (1887).
[3] Meyerhoffer, Monatsheft. Wien. **17**, 13 (1896).

present the only experimental investigation is one by Löwenherz.[1] The system studied was the one made up of potassium chloride, potassium sulfate, magnesium sulfate, magnesium chloride and water. The isotherm for 25° was determined, solution and vapor being always present. Magnesium sulfate and potassium chloride are the stable pair, and if no double salts were formed and no new hydrates there would be only two monovariant systems possible at this temperature, the solid phases being the stable pair of salts and potassium sulfate or magnesium chloride with six of water. It so happens that there are two double salts with compositions corresponding to the formula $K_2Mg(SO_4)_2,6H_2O$ and $KMgCl_3,6H_2O$. In addition we can have magnesium sulfate crystallizing with six of water and the number of monovariant systems becomes five, the solid phases being:

1. $K_2Mg(SO_4)_2,6H_2O$, KCl, K_2SO_4.

2. $MgSO_4,7H_2O$, $K_2Mg(SO_4)_2,6H_2O$, KCl.

3. $MgSO_4,7H_2O$, $MgSO_4,6H_2O$, KCl.

4. $MgSO_4,6H_2O$, KCl, $KMgCl_3,6H_2O$.

5. $MgSO_4,6H_2O$, $KMgCl_3,6H_2O$, $MgCl_2,6H_2O$.

It will clear matters up a little to consider the way in which we can pass from one monovariant system to another. Starting from a solution saturated with respect to potassium sulfate and potassium chloride, the following changes will take place on adding anhydrous magnesium chloride or on adding hydrated magnesium chloride and evaporating off the excess of water. Magnesium chloride and potassium sulfate go into solution, while potassium chloride is precipitated until the solution is saturated with respect to the double sulfate, and there are then three salts in equilibrium with the solution; potassium sulfate, potassium chloride, and potassium magnesium sulfate with six of water. With further addition of magnesium chloride, there is disappearance of potassium sulfate as solid phase and precipitation of the double salt and of potassium chloride, the concentration remaining unchanged until the solid potassium sulfate has vanished. The

---

[1] Zeit. phys. Chem. **13**, 459 (1894).

double sulfate will then dissolve and potassium chloride precipitate, the concentration changing until magnesium sulfate with seven of water appears, forming a new monovariant system. The double sulfate then disappears with formation of magnesium sulfate heptahydrate and potassium chloride. In the next stage, addition of magnesium chloride produces a slight precipitation of the two salts just mentioned, the amount of magnesium chloride increasing until the hexahydrate of magnesium sulfate separates. The heptahydrate will next disappear and the concentration will then change, the magnesium chloride precipitating both the hexahydrate and potassium chloride. The appearance of the double chloride is followed by the disappearance of the potassium chloride and then, with almost no change of concentration, by the appearance of crystallized magnesium chloride with six of water, forming the monovariant system marked V in the list. The data upon which these statements rest are given in Table XXXI. Departing from the usual custom the figures denote reacting weights of K_2, Mg, Cl_2 and SO_4 in one thousand reacting weights of water. The figures in the first column show the monovariant systems to which the concentrations refer.

TABLE XXXI

	K_2	Mg	Cl_2	SO_4
1	25	32	46	11
2	9	71	64	16
3	8	77	70	15
4	2	110	100	12
5	2	111	101	12

It is to be noticed that the two magnesium sulfates with seven and with six of water, can exist simultaneously in equilibrium with solution and vapor at a series of concentrations for each temperature. If we are to consider the appearance of the second salt as due to dehydration, the part of the isotherm along which these two hydrates are the solid phases must also be an isobar or line of constant pressure. There is nothing in the solubility determinations to vitiate this conclusion. Löwenherz finds that the two hydrates are in equilibrium with a solution containing fifteen reacting weights of magnesium sulfate and seventy-three reacting weights of magne ium chloride in one thousand reacting weights of water, while the concentration of the solution in equilibrium with the two hydrates and potassium chloride is given under 3 in Table XXXI. Assuming complete electrolytic dissociation there would be two hundred and forty-nine units in the first solution and two hundred and forty-eight in the second, numbers which are identical. This does not prove anything because the assumption is not fulfilled, and it is impossible to replace it by one which certainly represents the facts. Since it is not improbable that the degree of dissociation is much the same in the two solutions, this coincidence is sufficient to make an experimental study of this question very much to be desired.

ERRATA.

On page 19 it is stated that "at constant pressure the addition of heat produces an increase of volume" which is not true for liquid water between 0° and 4°. In the discussion on pages 10 and 26, it should therefore have been pointed out that the monovariant system, water and water vapor, will change, if cooled at constant volume, into the divariant system, liquid water, provided the ratio of the volumes of liquid and vapor be very large at 4°.

On page 30 strike out from "Although these results" to the end of the paragraph.

On page 44 I had forgotten a second paper by Hannay, Proc. Roy. Soc. 30, 484 (1880), in which he used a saturated solution.

In Fig. 5 the dotted lines should not appear to pass through a minimum pressure before meeting OB.

On page 177, instead of "To obtain the salt pure . . . ," read: The crystals will contain potassium chloride as impurity if washed with water or a potassium chloride solution and copper chloride as impurity if washed with a copper chloride solution.

INDEX OF AUTHORS

Alexejew, 69, 102, 103, 106, 127
Alkemade, van, 69, 147, 149, 166, 222.
Altschul, 14.
Amagat, 18, 20.
Ambronn, 39.
Andrews, 14.
Appleyard, 200.
Arctowski, 49.

Babo, v., 44.
Bailey, 50, 92.
Bancroft, 35, 36, 39, 43, 47, 69, 73, 81, 92, 98, 99, 101, 103, 127, 147, 159, 180, 240.
Bathrick, 159.
Battelli, 18.
Barus, 11.
Bauer, A., 124.
Bauer, A. E., 134.
Beckmann, 136.
Beilstein, 124.
Bemmelen, van, 200.
Berthelot, D., 31, 34.
Berthelot, M., 41, 239.
Bijlert, van, 136, 144.
Le Blanc, 39, 202.
Bodländer, 203, 241.
Boguski, 29.
Bois-Reymond, du, **138**.
Bosse, 212.
Braun, 4, 51, 52, 54.
Brodie, 29.
Brown, 118.
Budde, 82.

Cailletet, 14, 15, 25.
Carnelly, 42, 154.
Chancel, 41.
Chappuis, 140.
Le Chatelier, 1, 4, 30, 35, 41, 65, 76, 199.
Cohen, 57.
Colardeau, 14, 25.
Colson, 36, 52.
Coppet, de, 175.

Dahms, 129.
Dalton, 35.
Damien, 18.
Dammer, 31, 59.
Debray, 65.
Demarçay, 16.
Deventer, van, 42, 180, 181, 189, 190, 245.
Deville, 139.
Dewar, 18.
Dieterici, 44.
Ditte, 178.
Donny, 23.
Dufour, 23.

Duhem, 4, 22.

Emden, 44.
Étard, 41, 42, 44, 48, 49, 51, 128, 133, 134, 158.

Faraday, 15.
Favre, 52.
Ferche, 25.
Ferratini, 136.
Fock, 35, 36, 199, 203, 210, 212.
Frankenheim, 242.

Galitzine, 14, 35.
Garelli, 136.
Gautier, 143, 144, 145.
Gernez, 23, 32, 68.
Gibbs, 1, 2, 4, 22, 69, 100, 122, 147, 226, 234.
Gooch, 92.
Goldschmidt, 181, 187.
Gossens, 6.
Graham, 200.
Guldberg, 45.
Guthrie, 38, 40, 45, 48, 103, 107, 117, 125, 129, 133, 142, 153, 154, 157, 175.
Guye, 119.

Hagen, 143.
Hallock, 16.
Hannay, 36, 44, 97.
Hautefeuille, 33.
Haywood, 98.
Heide, van der, 165, 167, 170, 180, 216.
Heilborn, 15.
Helmholtz, H. v., **139**.
Helmholtz, R. v., 22.
Heycock, 142, **144**.
Hoff, van 't, **3**, 29, 35, 136, 139, 172, 173, 180, 181, 189, 190, 232, **245**.
Hoitsema, 140, 232.
Horstmann, **127**.

Isambert 65, 198.

Jakowkin, 239.
Joannis, 60, 65.
Jorissen, 181.
Jungfleisch, 239.

Konowalow, 59, 96, 98, **99**, **100**, **101**, 118, 119.
Kopp, 142.
Kuntze, 199, 203, 210.
Küster, 39, 136, 137, 144, 199, 200.
Landolt and Börnstein, 14, 40, 48, 154.
Lang, v., 20.
Lehfeldt, 125.
Lehmann, 23, 32, 33, 34, 93, 129, 133, 134, 137, 138, 143, 242.
Lescoeur, 60.

Linebarger, 85, 119, 125.
Link, 241.
Löwenherz, 186, 246, 248.

MacGregor, 52.
Mallard, 30.
Marsden, 36.
Margueritte-Delacharlonnay, 92.
Masson, 103.
McIntosh, 237.
Meslans, 31, 33, 140, 145.
Meyerhoffer, 22, 33, 43, 146, 169, 170, 174, 175, 177, 178, 180, 181, 206, 209, 213, 215, 216, 231, 245.
Miolati, 116, 129.
Mitscherlich, 31.
Mond, 140.
Morse, 92.
Moser, 23.
Muthmann, 199, 203, 210.

Nadeshdin, 14.
Natterer, 15.
Nernst, 2, 5, 21, 27, 35, 46, 92, 93, 100, 103, 108, 234, 239.
Neuberg, 241.
Neville, 142, 144.
Nicol, 158, 159, 202, 203.
Noyes, 202.

Oersted, 20.
Offer, 38.
Olszewski, 15.
Orndorff, 124.
Ostwald, 18, 20, 23, 35, 42, 44, 51, 52, 58, 63, 68, 92, 97, 98, 99, 122, 124, 127, 129, 130, 138, 142, 166, 200, 201, 235, 242.

Pagliani, 20.
Parmentier, 41.
Paternò, 136.
Pfaundler, 38.
Pfeffer, 235, 236.
Pfeiffer, 240.
Pickering, 107, 113, 114, 131, 132, 220.
Pictet, 15.
Prendel, 28.
Prytz, 91.

Ramsay, 12, 17, 25, 140.
Raoult, 44, 236.
Reicher, 29, 30, 31, 32, 42, 91, 179, 180, 181, 245.
Remsen, 38.
Retgers, 207.
Riecke, 31.
Roberts-Austen, 36, 162.
Roloff, 50, 129.
Röntgen, 20.
Rose, 33.
Roscoe, 122.

Roozeboom, 3, 24, 30, 31, 35, 41, 47, 43, 61, 63, 65, 69, 71, 75, 76, 78, 80, 81, 82, 83, 85, 91, 98, 106, 107, 109, 111, 112, 113, 114, 137, 140, 147, 149, 152, 153, 155, 162, 163, 168, 169, 172, 174, 179, 188, 189, 190, 199, 201, 204, 205, 206, 208, 209, 212, 217, 219, 221, 222, 223, 226, 231, 232.
Ruys, 31.
Sajoutchewsky, 14.
Schmidt, C., 241.
Schmidt, G. C., 200.
Schneider, 20, 36.
Schreinemakers, 146, 147, 150, 159, 166, 168, 178, 183, 196, 201, 202, 205, 213, 214, 219, 223.
Schröder, 23.
Schrotter, 23, 33.
Schumann, 20.
Schultz, 38, 130, 143.
Schützenberger, 93.
Schwarz, 34, 133.
Shields, 140.
Shenstone, 41, 129.
Spring, 16, 21, 133, 189, 232.
Stackelberg, v., 51.
Stock, 136.
Stokes, 147.
Stortenbeker, 35, 86, 89, 199, 201, 204, 207, 209, 212.
Story-Maskelyne, 36.

Tammann, 142, 144.
Tegetmeier, 36.
Thoma, 139.
Thomson, 18.
Thorpe, 118.
Thurston, 147.
Tilden, 41, 129.
Traube, J., 241.
Traube, M., 235.
Trevor, 4, 100, 205, 212, 226.
Troost, 33.

Valson, 52.
Vaubel, 42.
Vicentini, 20.
Vignon, 129.
Villard, 92.
Violle, 36.
Voigt, 18.
Vriens, 194.

Waals, van der, 14, 23, 89.
Wald, 25, 226.
Walden, 235.
Walker, 96, 200.
Warburg, 36.
White, 92.
Winkelmann, 45, 125.
Wroblewski, 15.

Young, 12, 17, 25.

INDEX

Absorption, 138.
Adsorption, 138, 200.
Acid, benzoic, 105.
  formic, 118, 122.
  hydrobromic, 91, 98, 112, 118, 122.
  hydrochloric, 91, 114, 118, 122, 229.
  salicylic, 105, 242.
  titanic, 33.
Alcohol, butyl, 103.
  ethyl, 45, 127, 159, 240, 241.
  isobutyl, 103.
  methyl, 92, 127, 236.
  propyl, 118, 121, 123, 127.
Allotropic modifications, 28, 31, 33, 34, 71, 87, 93, 129, 133, 145.
Alloy, 142.
  eutectic, 117, 129, 149, 156.
Ammonia, and ammonium bromide, 78.
  of crystallization, 66.
  and sodium, 61, 65.
Ammonium chloride, 17, 181, 198, 201, 206, 212, 217, 227, 242, 244, 245.
Antimony, 144, 145.
Arragonite, stability of, 33.

Barium carbonate, dissociation of, 65.
Boiling point, 12.
  change with pressure, 13.
  of divariant systems, 50, 101, 120.
  of monovariant systems, 12, 43, 50, 87, 95, 101.
Break, 67, 128, 132, 152, 159, 169, 240.

Calcite, stability of, 33.
Calcium carbonate, dissociation of, 65.
Calcium acetate, 179, 181, 188.
  butyrate, 41.
  isobutyrate, 41.
  chloride, 71.
  sulfate, 41.
Capillary tension, 2, 8.
Carbon, stability of modifications of, 33.
Chlorine, 86, 112.
Component, 1, 226, 227, 234.
Compound, 1, 35, 143.
  chemical and molecular, 56.
Consolute, 35.
  temperature, 103, 127, 143.
Constituent, 226, 227.
Copper, acetate, 179, 181, 188.
  chloride, 175, 180, 181, 191, 194, 201, 212, 213, 215, 216.
  sulfate, 201, 202, 204, 205, 207, 208, 212, 236.

Critical, pressure, 14.
  temperature, 14, 18, 44.
Cryohydrate, 39.
Cryohydric temperature, 39, 117, 129, 157, 178.
Curve, boundary, 25, 149.
  fusion, 45, 91, 94, 96, 106, 117, 128, 132.
  middle, 152.
  side, 152.
  solubility, 45, 104, 105, 187.
  sublimation, 25.
  vaporization, 24.

Deliquescence, 51, 64.
Density relations, 7.
Diagrams, 24, 66, 69, 99, 173, 146, 159, 170, 202, 209, 231.
  concentration-pressure, 96, 118, 128.
  concentration-temperature, 66, 70, 73, 79, 85, 88, 94, 101, 105, 106, 110, 116, 120, 126, 131, 133, 135.
  pressure-temperature, 24, 26, 32, 38, 46, 57, 71, 72, 73, 87, 108, 112, 173, 184, 186, 188, 190, 192.
  triangular, 148, 154, 157, 161, 165, 172, 176, 193, 220.
Diaphragm, semipermeable, 235, 236, 237.
Diethylamine, 103, 132.
Dissociation, 65, 187, 189, 197.
Distillation, of consolute liquids, 120.
  of partially miscible liquids, 102.
Divariant, 3.
  systems, 3, 4, 37, 146, 243.

Efflorescence, 59, 62, 64, 186, 193.
Equilibrium, complete heterogeneous, 3.
  incomplete heterogeneous, 3.
  labile, 22.
  stable, 22.
Eutectic, alloy, 117, 129, 149, 156.
  temperature, 117, 129.

Ferric chloride, 78, 201, 206, 217, 219.
Formula, use of, 229.
Fractional, crystallization, 218.
  distillation, 102, 120, 124.
  evaporation, 209.
Freedom, degrees of, 3, 230.
Freezing mixtures, 47.
Freezing point, 7.
  change with pressure, 18, 49, 91.
  of solutions, 46, 142.
  of solid solutions, 135, 144.

Fusion curve, 45, 91, 94, 96, 106, 117, 128, 132.
Fusion point of two solid phases, 163, 222.

Gravity, effect of, 2, 5, 13, 20, 100.

Heat, addition and subtraction of, 8, 12, 16, 17, 19, 27, 54, 232.
relations, 7.
reservoir, 9.
of solution, 42, 102, 103.

Hydrate, 56.
melting point of, 64, 72, 79, 112, 131, 219, 221.
vapor pressure of, 59, 73, 81, 186, 189, 194.

Inversion, point, 25, 57, 75, 93, 180, 233, 239, 243.
pressure, 3, 6, 25, 30.
temperature, 3, 6, 25, 29, 30, 33, 61, 133, 180, 238, 239.

Iodine, 86, 92, 136, 199.

Isotherm, 155, 156, 161, 201, 209, 213, 223, 240, 241, 246.

Labile, 22.

Lead iodide, 179, 195, 205, 213.
nitrate, 153, 202.

Liquefaction of gases, 15.

Liquids, properties of, 20.
supercooled, 23.
superheated, 24.

Magnesium, chloride, 245, 246.
sulfate, 165, 172, 180, 190, 191, 195, 201, 202, 205, 207, 212, 213, 216, 240, 242, 245, 246.

Manganous, chloride, 212.
sulfate, 85, 208, 240, 242.

Matter, existence of, 5.

Mercuric iodide, 134, 151.
sulfate, 230, 232.

Metals, 142.
diffusion of, 21.
volatility of, 16.

Monovariant, 3.
systems, 3, 4, 37, 146, 235, 237.

Naphthalene, 6, 18, 25, 44, 94, 104, 116.

Nonvariant, 3.
systems, 3, 4, 37, 146, 235, 237, 243.

Occlusion, 138.

Palladium and hydrogen, 139.
Pattinson process, 142.
Phase, 1.
Phenanthrene, 116.
Phenol, 103, 105, 136.
Phosphorus, 32.
Potassium, chlorate, 133, 137, 201, 204, 207, 212.
chloride, 39, 157, 175, 180, 191, 194, 201, 212, 213, 215, 216, 228, 231, 244, 245, 246.

Potassium, hydroxide, 131, 227, 240.
iodide, 44, 92, 179, 195, 205, 213.
nitrate, 129, 133, 153, 157, 201, 202, 207, 227, 228, 231, 243, 244.
perchlorate, 201, 203, 210.
permanganate, 201, 203, 210.
sulfate, 165, 180, 191, 201, 203, 205, 207, 210, 212, 216, 230, 244.

Potential, chemical, 2.

Pressure, and boiling point, 13.
critical, 14.
and freezing point, 18, 49, 91.
inversion, 3, 6, 25, 30.
osmotic, 235, 236.
and solubility, 51.
vapor, 11, 16, 43, 50, 59, 73, 81, 87, 91, 96, 99, 118, 185, 186, 189, 194, 198, 248.

Racemates, 181, 233.
Range of decomposition, 174.
Reaction, chemical and physical, 5.
velocity of, 25, 31, 32, 85.
Resistance, passive, 226, 234.

Silver, iodide, 134, 151.
nitrate, 128, 133, 201, 207.

Sodium, and ammonia, 61, 65.
chloride, 54, 159, 184, 201, 210, 228, 231, 243, 244, 245.
nitrate, 128, 153, 201, 202, 210, 231, 244.
sulfate, 41, 56, 172, 180, 184, 190, 195, 201, 205, 207, 212, 213, 232, 240, 242, 245.

Solids, properties of, 21.
volatility of, 92.

Solubility, change with pressure, 51.
change with temperature, 41.
and chemical nature, 42.
of consolute liquids, 127, 240.
curve, 45, 104, 105, 187.

Solute, 35, 42, 94, 109, 147, 154, 157.
change of solute, 89, 106, 128, 133.

Solution, 35.
heat of, 42, 102, 183.
solid, 35, 135, 144, 199, 201, 204, 239.
three liquid, 239.
two liquid, 94, 96, 99, 101, 103, 107, 112, 238, 240, 241, 242.
supersaturated, 68, 81, 221.
vapor pressure of, 43, 50, 87, 91, 96, 99, 118, 185, 194.

Solvent, 35, 43, 94, 109, 147, 154, 157, 183.
change of, 89, 106, 128, 133.

Sublimation, curve, 25, 186.
point, 17.

Sulfur, 28.
and toluene, 44, 93, 98, 103.
and xylene, 98, 102.

Sulfur dioxide, 98, 107.

Tartrates, 181, 233.
Temperature, consolute, 103, 127, 143.
  critical, 14, 18, 44.
  cryohydric, 39, 117, 129, 157, 178.
  eutectic, 117, 129.
  inversion, 3, 6, 25, 29, 30, 33, 61, 153, 180, 238, 239.
Thallium chlorate, 157, 201, 204, 212.
Thorium sulfate, 41, 85.
Toluene, 93, 98, 103.
Transformation point for two solid phases, 223.
Triethylamine, 106.

Vapor, and gas, 15.
  properties of, 19.

Vapor, pressure, 11, 16, 198.
  pressures of hydrates, 59, 75, 81, 186, 189, 194, 248.
  pressure of solutions, 43, 50, 87, 91, 96, 99, 118, 185, 194.
  supercooled, 7, 22.
Vaporization curve, 24.
Variable, dependent, 226.
  independent, 2, 226, 235.
Volatility, of metals, 16
  of solids, 92.

Water, 16, 24, 44.
  of crystallization, 56, 201.
Work, addition and subtraction of, 8, 11, 16, 19, 27, 55.

www.ingramcontent.com/pod-product-compliance
Lightning Source LLC
Chambersburg PA
CBHW021354230426
43666CB00006B/522